HOW THINGS WORK

HOW THINGS WORK

EVERYDAY
TECHNOLOGY
EXPLAINED

BY JOHN LANGONE

ART BY PETE SAMEK,
ANDY CHRISTIE,
AND BRYAN CHRISTIE

NATIONAL GEOGRAPHIC

WASHINGTON, D.C.

CONTENTS

Preceding pages: Interlocking gears, simple yet effective, keep many things operating like clockwork.

INTRODUCTION

T HAS BEEN SAID THAT WHENEVER THOMAS Edison showed visitors the numerous inventions and gadgets that filled his home, someone would invariably ask why they still had to enter through an old-fashioned turnstile. The master inventor had an easy and altogether practical answer: "Because every single soul who forces his way through that old stile pumps three gallons of water up from my well and into my water tank."

Versatility, as well as the ability to perform useful work, is what machines and mechanisms are often all about, and while a device may have a specific use, chances are that its operating principles and many of its parts, if not the device itself, can be put to work in a variety of ways. The compact disc, for example, fills our homes with music, but it also can provide the images, data, and sound that emerge from a personal computer. Propellers drive planes through the air, but they also help generate electricity in a hydroelectric power plant. Water irrigates soil-grown crops, and it can itself become the medium for growth on a soilless hydroponics farm. Fabrics

let air in but also keep it out; lenses allow us to look deep inside our bodies and far into the universe. An electron microscope works more like a television set than like a conventional microscope; the telephone expands into the Internet; the steam engine lends a hand to drive a nuclear-powered submarine; the levers that move a piano's keys are akin to those that run a typewriter.

The machines and other technologies that have extended the range of human capability are all products of a human ingenuity that has successfully utilized the science behind motion, forces, and various forms of energy. To truly understand how a machine works—and a machine can be something as simple as a lever, an inclined plane, a wheel and axle, a pulley, or a screw—one has to have at least a casual acquaintance with nature's forces and how they are modified, transformed, transmitted, and otherwise adapted to do work.

In general terms, machines work because some kind of force, or energy that affects motion, is applied. Put another way, machines are devices

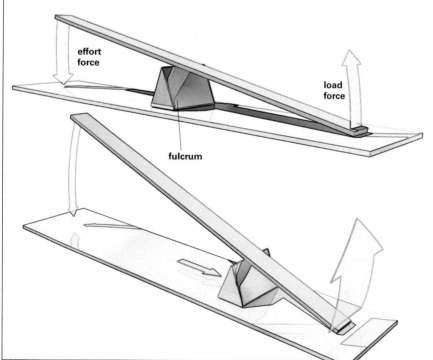

effort force

load force

fulcrum

LEVERS
These simplest of mechanical contrivances require effort at one end to raise a load on the other, and the effort required to lift the load depends on the position of the fulcrum. With equal amounts of effort and load at its ends, a lever balances when its fulcrum occupies a position at the center. Moving the fulcrum farther from the load (top) makes the load harder to raise. Placing it closer to the load (bottom) reduces the amount of effort needed for lifting.

PULLEYS

(Below) Another basic machine, a pulley may consist of grooved wheels, mounted on blocks, with a rope running in the grooves. A fixed pulley changes the direction of force applied to a rope. If you run the end of that rope around an unfixed pulley and then pass it back to the fixed one, you can raise a load attached to the rope with half the effort. Various combinations of ropes and pulleys help distribute the weight of a load, easing the strain on workers who must lift heavy objects.

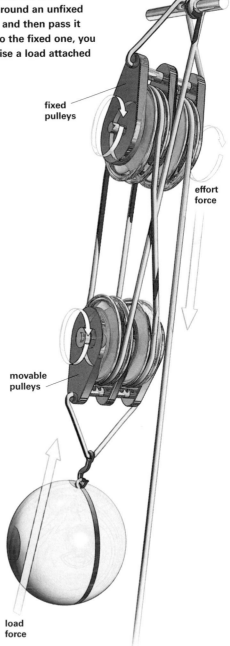

fixed pulleys

effort force

movable pulleys

load force

that overcome resistance at one point by the application of a force at some other point. Various movable mechanical parts, such as levers, gears, and springs—or electrical wires, transistors, and electromagnets—manipulate and transmit the force exerted by a "prime mover," sending it where it is needed and getting the most out of it. This is where efficiency comes in.

Always less than 100 percent, since some energy is lost to friction as heat, a machine's efficiency is the ratio of useful work to the energy put into it, and it varies enormously depending on the machine. Only about 6 to 8 percent of the energy from burned coal went into the work that the old steam locomotive did to haul the train. On the other hand, a gasoline engine and a steam turbine may have a thermal efficiency of around 25 percent, a diesel oil burner about 35 percent, a hoist some 60 percent, and an electric motor around 95 percent.

The process through which a machine works often involves interconnections: As a force moves one component of a machine, that part exerts force at another point, moving some other part, and so on. Thus energy is handed along by the machine in a sort of relay. In a nuclear power plant, for example, the force of heat from split atoms vaporizes water, creating the steam that turns the turbine that runs the electrical generator. In an automobile, vaporized gasoline ignites and burns inside the cylinders, transmitting motion to the wheels.

But the laws of force and motion do not apply only to running a steam engine or driving a vehicle. In the human body, chemical energy moves the muscles that do the work when we swing axes or hit golf balls, while the bones act as levers, and the joints are controlled by the body's equivalent of belts and pulleys. Fired bullets and rockets follow the force-motion rules. The water that rushes from a faucet, the furnace that forces heat into a room, and the electrical pulses that actuate a loudspeaker do so as well.

The simplest of our machines is probably the lever: a rigid beam or rod pivoted at a fulcrum point. Used in one basic form for centuries, at first to lift heavy stones or tree trunks, it is now used in crowbars, wheelbarrows, nutcrackers, computer keyboards, pliers, wrenches, nail clippers, kitchen scales, and even the links within

a piano that move the hammers that strike the tightened wires to sound the notes.

A pulley, which is a wheel—another simple machine—on a shaft, is also a type of lever, as are the gear wheels of a clock, which form a system of levers whose lengths are the radii of the wheels. The inclined plane is a simple machine that enabled the ancient Egyptians to haul up stones to build the pyramids; its spiral cousins include the screw and the automobile jack.

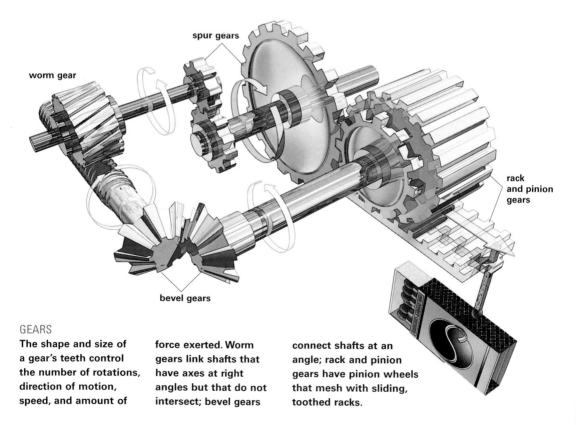

spur gears

worm gear

rack and pinion gears

bevel gears

GEARS

The shape and size of a gear's teeth control the number of rotations, direction of motion, speed, and amount of force exerted. Worm gears link shafts that have axes at right angles but that do not intersect; bevel gears connect shafts at an angle; rack and pinion gears have pinion wheels that mesh with sliding, toothed racks.

The operating principle of an inclined plane (effort to move an object to a higher plane is reduced by extending the distance the object must travel) also governs the action of the wedge, which translates work at one end to force at the sides, as in the action of splitting wood.

Last, but certainly not least, are wheels and axles, which take advantage of the principle that rolling friction is less of a drag than sliding friction. Shoving a heavy box across a floor, for example, may be impossible without rollers of some kind beneath; with them, however, friction is dissipated and relatively little force is required. A power and motion transmitter par excellence, the wheel mediates between rotary and linear motion, stores energy—as in a flywheel—and shows up in gears, winches, capstans, turbines, and pulleys.

The mechanisms and principles behind simple machines, and the complex ones that they create when mixed and matched, are what this book is all about. *How Things Work* is not a home repair book, though. It will not tell you how to fix a cell phone or tune a piano, for example. Nor is it a manual that will teach you how to pilot a plane or stitch a hem with a sewing machine. What it will do is satisfy your curiosity about how and why these machines work. It will also tell you about a host of technologies that affect our lives daily.

The pages ahead include more than information on conventional machinery. Other technologies—those that appear to be "non-machines," for want of a better phrase—are equally interesting, and while they are not gadgets in the classic sense, they still have a place here because they, too, rely on scientific principles to function. Thus, the book not only explores what goes on inside a refrigerator, a computer, a jet engine, and other familiar machines, but it also examines the technology and the principles behind achievements

A MOTOR

A motor converts electrical into mechanical energy—just the reverse of what a generator, or dynamo, does—and it works on the principle that an electrical conductor moves when in the presence of a magnetic field at right angles to the current. Current enters the motor via a commutator and its brushes, and it turns a laminated steel rotor that contains conducting coils. The commutator reverses the flow of current every half turn of the coils, and this action reverses the magnetic field, thus ensuring that the coil continues moving. As the rotor turns, a shaft delivers mechanical power sufficient to drive mixers, blenders, can openers, and many other household appliances.

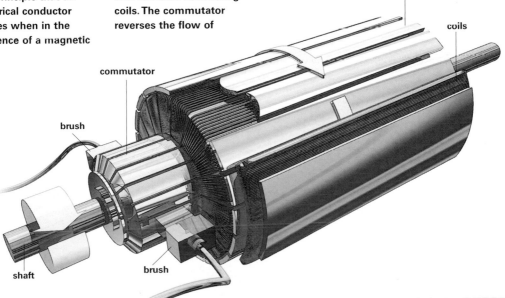

rotor

coils

commutator

brush

brush

shaft

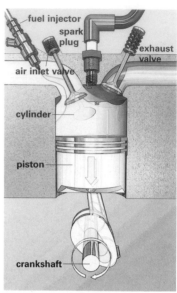

POWER STROKES

(Right) In a four-stroke internal combustion engine, vaporized gasoline ignites and burns inside the cylinders. The downward induction stroke opens an inlet valve, sucking in fuel and air; on compression, an upward stroke compresses the air and fuel mixture; in the downward power stroke, a spark plug ignites the fuel; on exhaust, a valve opens, expelling gases. Variations of the internal combustion engine can power a wide variety of things, from lawn mowers and leaf blowers to bulldozers, tanks, and submarines.

Induction

such as building arches, genetically altering animals and crops, designing fireworks displays, and synthesizing detergents that clean.

A computer, logic gates, and the Internet are stripped to their bare essentials. The mysterious mechanism behind a clock's face and the powerhouse inside a nuclear submarine are opened to view. From nylon stockings to a construction crane, from a maglev train to the stock exchange, everyday technology is not only explained but also rendered easy on the eyes.

Within each chapter, sidebars give you insight into such topics as the pendulum, fish farming, satellite radio, and hybrid-powered automobiles. They also help to highlight the application of science to specific technologies and will provide a grace note or two in the midst of some of the more technical discussions. You can open this book at any place and read each spread as if it were a magazine article.

Moreover, because technology overlaps so many fields, the interdependence of the scientific, industrial, and technical disciplines involved in creating and utilizing our machines, equipment, and systems is always hovering implicitly or stated explicitly in the text. The overlap becomes especially obvious in the section describing the MRI scanner, which is a wedding of medicine, electronics, computer technology, and nuclear physics. You can also see it in the discussions of such industrial processes as the manufacture of steel, plastics, and glass, all of which involve the blending of basic science and technology.

Inventions and innovations are, of course, the creations of the human mind. Where possible, *How Things Work* includes historical information about a gadget or a process, relating how, when, and why it came to be and describing its impact on society. We learn, for instance, that clocks may have been in use in 2000 B.C., that Leonardo da Vinci designed a steam-powered vehicle, that Queen Elizabeth I refused to grant a patent to the inventor of a knitting machine, and that miners were once paid by the tons of coal they loaded by hand.

The book's focus on technology should not, however, be construed as our endowing its intricate and awe-inspiring workings with the same divine right as kings, or as our implying that

Compression

Ignition

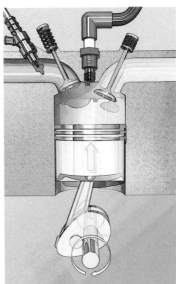
Exhaust

it is omnipotent and infallible. Red flags certainly abound. Computerization, our new symbol for automation and mechanization, has become, for some people, the cyber-devil that drains away human reasoning and dumbs us down, makes us slaves to speed and materialism, and increasingly isolates us.

Sperm and eggs are now routinely manipulated outside their natural hosts. The human genome has been thoroughly mapped. Embryos are transplanted, sheep and other animals are cloned. All these developments of science and technology raise significant moral and ethical questions. A generation that knows almost no other way relies on a digital language to perform countless tasks that once took not only hands-on effort but also human innovation, intuition, and imagination. Military technology—with its almost unbelievably expert arsenal of "smart" weapons and systems—has a purpose that the civilized world would rather not have to boast about.

In light of these caveats, it is no wonder that science and technology have had their detractors. Mark Twain, in acknowledging that science was fascinating, mused, "One gets such wholesome returns of conjectures out of such trifling investment of fact."

For George Bernard Shaw, science was always wrong. "It never solves a problem without creating ten new ones," he said. For Oliver Wendell Holmes, it was "a first rate piece of furniture for a man's upper chamber, if he has common sense on the ground floor."

More to the point may be a remark by Canadian physician Wilder Penfield. "The trouble is not science," Penfield observed, "but in the uses men make of it. Doctors and laymen alike must learn wisdom in their employment of science, whether this applies to atom bombs or blood transfusions."

Perhaps science and technology will never solve all of nature's mysteries. Maybe they have given us too much already, including some things that are not that desirable. But in the balance, as this book will show, it's a world of wonder that we have created with our human history of tools and ingenuity. In fact, without today's science and technology, we would have very little at all.

AT HOME

B E IT EVER SO HUMBLE, THERE IS NO PLACE LIKE a modern home. Furnaces, fans, refrigerators, stoves, washing machines—today's comfortable, efficient structures are pierced by a circuitous system of wires, pipes, vents, and ducts. They rely on electricity and the transformer, a 19th-century invention that made it possible to transmit power for domestic and industrial use. Because we have electricity, houses are filled with "home basics" people once considered luxuries. Some homes have digitally connected environments that link television sets, personal computers, programmable thermostats, lighting, and security systems. Eventually, most of the pieces of equipment in a home will be linked electronically in a similar network. The average home contains a hundred pieces of powered equipment—machines that can reduce our physical labors and free up time.

Just as suburban houses form a tidy network of parts, so the appliances and systems in each one of them fit together to make modern life more comfortable.

COOKING

N 1900, THE SEARS, ROEBUCK AND COMPANY CATALOG OFFERED A top-of-the-line Acme range that burned coal or wood and sold for a whopping $31.05. Today a high-tech home cookstove with computerized controls can cost thousands of dollars, but a person doesn't have to pay anywhere near that to cook like Julia Child.

Standard electric and gas ranges are efficient heat producers, while microwave ovens, toaster ovens, electric skillets, and slow cookers are good alternatives. Each has its own size, shape, and purpose, and all but the microwave convert electricity into heat. As electricity passes through insulated wires inside metal coils or loops, called heating elements, the electrical resistance heats the outer metal. Controls regulate the heat by adjusting voltage in the wires, or a thermostat makes adjustments automatically.

In a gas range, natural or bottled gas flows in and is mixed with air in a chamber; then the mixture is ignited by a spark, the flame of a pilot light, or an electric heating coil, firing the burners. In the convection oven, a fan circulates heated air uniformly and continuously around food for faster and more even cooking. The induction range—also called cool electric induction—uses a magnetic field, generated electronically, under a ceramic range-top. Instant heat is generated through resistance when magnetic pans are set on the ceramic plate.

A microwave oven produces high-frequency electromagnetic waves. Passing through food, the waves reverse polarity billions of times a second. The food's water molecules react to each change by rapidly reversing themselves. Friction results, heating the water and cooking the food.

SUSPENDED ANIMATION
(Right) Heat, latches, timers, and springs pop toasted bread out of a toaster.

MICROWAVE OVEN
(Below) Microwave ovens use the same electro-magnetic radiation as radar does to magnetically agitate water molecules in food (below left), causing them to heat faster than in a conventional oven. Produced electrically in a magnetron (below right), the high-frequency microwaves pass through a wave-guide, encounter a stirrer-fan, and reflect into the oven. Invisible and seemingly benign, microwaves can injure human tissue.

SEE ALSO
Home Heating & Cooling · 22
Sending Signals · 118

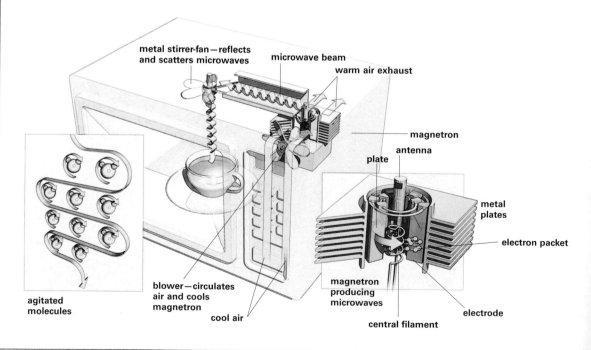

metal stirrer-fan—reflects and scatters microwaves
microwave beam
warm air exhaust
magnetron
antenna
plate
metal plates
electron packet
agitated molecules
blower—circulates air and cools magnetron
cool air
magnetron producing microwaves
central filament
electrode

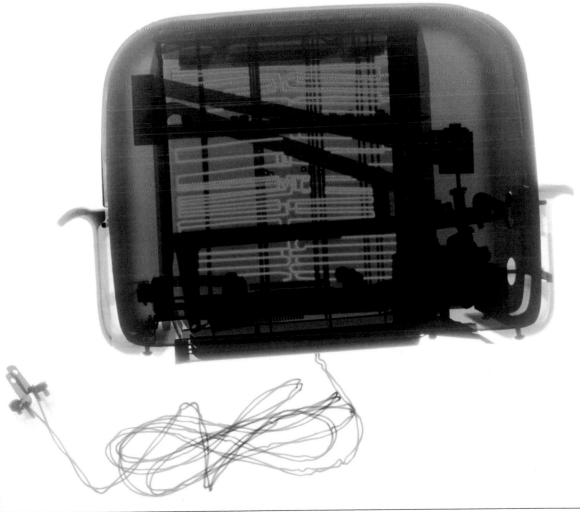

REFRIGERATORS

I N THE DAYS OF THE EARLY ICEBOX, THE SIMPLE PROCESS OF ICE melting inside the enclosure took up a certain amount of heat and kept the icebox cool. If someone covered the block of ice with paper to conserve the coolant, the box would not function properly because the paper kept the ice from melting. The icebox was quickly replaced when the first Frigidaire came off a 1921 assembly line at the Delco Light Plant, a subsidiary of General Motors.

The modern refrigerator also cools by extracting heat, but in a more complicated way. A compressor pumps coolant vapor through sealed tubes; after increasing its pressure and temperature, it routes the vapor outside the box and into a condenser, where the coolant releases heat and becomes a liquid. From there, pipes lead the liquid back into the box, through a control valve, and into the evaporator, or freezing unit; the coolant vaporizes in the coils surrounding the unit and absorbs heat, thereby cooling the unit. The warmed vapor returns to the compressor, and the cycle begins again.

A freezer compartment stays at a temperature typically of 0°F to 10°F, cold enough to preserve food up to a year, and the refrigerator enclosure itself usually stays between 32°F and 40°F. The cold air checks the growth of bacteria and molds and slows the chemical breakdown of food.

Modern refrigerators often include automatic icemakers, which replace the trays that a homemaker used to fill with water to freeze into ice cubes. All that an icemaker generally needs to go through its cycle and make cubes is an electric motor and heating unit, a connection to a water line, a water valve to fill plastic molds, and a thermostat to regulate the water in the molds.

When the molds are filled and frozen, a heating coil warms their underside, and ejector blades attached to the motor free the cubes and spew them into a bin.

COOL COILS

(Right) Refrigerators depend on circulating a chemical refrigerant able to absorb and then release heat effectively. Refrigerated food is kept from deteriorating by chilled air, following a principle recognized long ago when people stored foodstuffs in cool caves. Cool air slows the growth of bacteria but, contrary to popular belief, does not kill them.

SEE ALSO

Home Heating & Cooling · 22

CHANGING CHILLERS

Scientific findings beginning in the 1970s suggested that the chlorofluorocarbons (CFCs) used in refrigerants—and in other common items such as aerosol cans and cleaning chemicals—released too much chlorine into the atmosphere. Such large amounts of chlorine gas, these studies suggested, were threatening the ozone layer, important to Earth's atmosphere because it absorbs harmful ultra-violet radiation beaming down from the sun. In 1987 nations around the world agreed to phase out the manufacture and use of all CFC-based refrigerants. Newly designed refrigerants have come on the market since, and today's refrigerators cool with these instead.

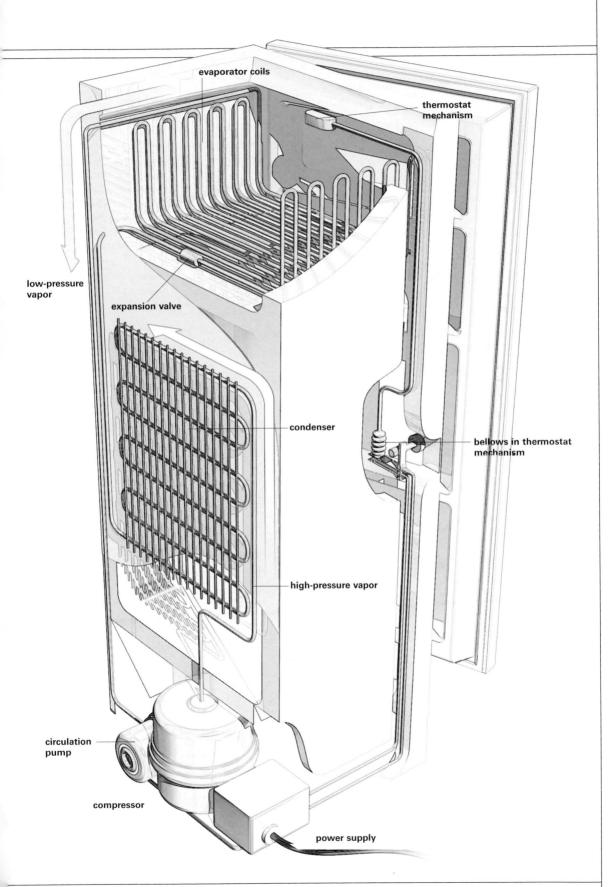

evaporator coils

thermostat mechanism

low-pressure vapor

expansion valve

condenser

bellows in thermostat mechanism

high-pressure vapor

circulation pump

compressor

power supply

VACUUM CLEANERS

THE SIMPLE CARPET SWEEPER PICKS LINT AND CRUMBS OFF FLOORS and rugs. Into an enclosed pan on wheels at the end of a long stick, small bits of debris are scooped up by the roller action of a cylindrical brush mounted on the underside. It's not much better than a dustpan and brush, but it does not require stooping or bending.

Electric vacuum cleaners, which made their debut early in the 20th century, use suction, not gathering, to pick up particles. A motor-driven fan creates a partial vacuum that sucks up dirt loosened by a beater brush and deposits it in a bag or other type of collector. Vacuum cleaner fan blades spin up to 18,000 times a minute; jet engine blades reach 8,000.

A central vacuum cleaner has a power unit and a collection canister in the basement, garage, or spare room that is connected to tubes in walls and under floors. To activate the system, a cleaning hose and its attachment-tool are plugged into inlet valves in the walls. Dust and debris are sucked through the tubes and sent to the dirt-collection canister, which has to be emptied a few times a year.

A new variation on vacuum cleaner technology are the battery-operated robot cleaners that scoot over floors and under beds and tables without having to be pushed or plugged in. Resembling a big CD player, they are equipped with wheels, rotating brushes and sweepers, bumpers, and small vacuum systems. Their computerized navigation systems and infrared or ultrasonic sensors keep them away from obstacles, including pets, and from toppling off stairs. Powered by nickel-hydride batteries, robot cleaners can be confined to specific areas by magnetic strips placed near open doorways or by electronic devices that set up invisible "walls."

HOUSEHOLD DUST
(Right) A vacuum cleaner may remove it from sight, but through an electron microscope, dust appears as a formidable opponent. This batch has cat fur, twisted synthetic and woolen fibers, serrated insect scales, a pollen grain, and plant remains. Loaded with an assortment of substances, simple household grit can cause asthma and other allergic reactions.

IT'S IN THE BAG
(Left) In its scientific sense, a perfect vacuum means a complete absence of matter within a space—a rare, if not impossible, phenomenon. In a more familiar sense, a perfect vacuum describes a powerful cleaning machine that removes a great deal of matter from our rugs. An upright unit uses a partial vacuum that fills a bag with so much dirt that it requires frequent emptying. Air blown through the unit by a motor-driven fan escapes through the bag's porous walls, but dirt remains trapped inside.

SEE ALSO

dust bag

motor

beater brush

fan

drive belt

WASHERS & DRYERS

TAKE-OUT LAUNDRY
(Left) Coin-operated and self-service, today's laundromat traces its ancestry to the nation's first: the "Washeteria," which opened in Fort Worth, Texas, in 1936.

TAKING IT FROM THE TOP
(Right, above) A cone-shaped, motor-driven agitator in a top-loading washer moves clothes back and forth in a tub filled with soapy water. When the wash cycle ends, a timer directs the motor to free the agitator and spin a perforated inner basket. Centrifugal force spins water through the holes into the outer tub and presses the clothes against the sides of the basket. After the water drains out, valves open and refill the tub for the rinse cycle. A final spin damp-dries the laundry. Springs attached to the tub and the unit's frame keep the washer stable during vibrations.

DETERGENT
(Right, below) A detergent molecule has one polar and one nonpolar end. The nonpolar end attaches to oily dirt while the polar end dissolves in the surrounding water, ionizing to form positive charges. These charges repel the ones on other soap-encrusted dirt particles, keeping the dirt suspended and allowing it to be rinsed away.

BEFORE THE WASHING MACHINE WAS INVENTED, PEOPLE USED washboards to scrub, or they carried their laundry to riverbanks and streams, where they beat and rubbed it against rocks. Such back-breaking labor is still commonplace in parts of the world, but for most home owners the work is now done by a machine that regulates water temperature, fills, washes, rinses, spin-dries, and empties automatically. With its intricate electrical, mechanical, and plumbing systems, the washing machine is one of the most technologically advanced large household appliances. It not only cleans clothes, but it does so with less detergent and energy than washing by hand requires. Compared with the old wringer-type washers that squeezed out excess water by feeding clothes through rollers—and emptied out wastewater through a hose—modern washers are indeed an electrical-mechanical phenomenon.

Manufacturers produce two types of washing machines: top-loaders and front-loaders. The top-loading model, which is generally more popular, relies on a motor-driven agitator unit that wraps and twists clothes around during the wash and rinse cycles. Pumps circulate, recirculate, and discharge water, while timers, switches, and sensors regulate the flow and temperature of the water, as well as the spin process. Some top-loaders can use an average of 40 gallons of water for each load of wash.

Front-loading machines are built without agitators. Instead of dragging and shoving clothes through a cycle, these machines churn the laundry about in high-speed rotating drums. According to their manufacturers, front-loaders use 20 to 25 gallons of water for each wash load.

The washer's companion is the dryer—electric or gas-powered—and, like electric and gas furnaces, this machine circulates the heat from coils or burners to do its job. A fan draws air into the dryer, where it is heated, passed through spin-damp clothes, and forced through a lint trap. The air is then ejected through an exhaust vent.

SEE ALSO

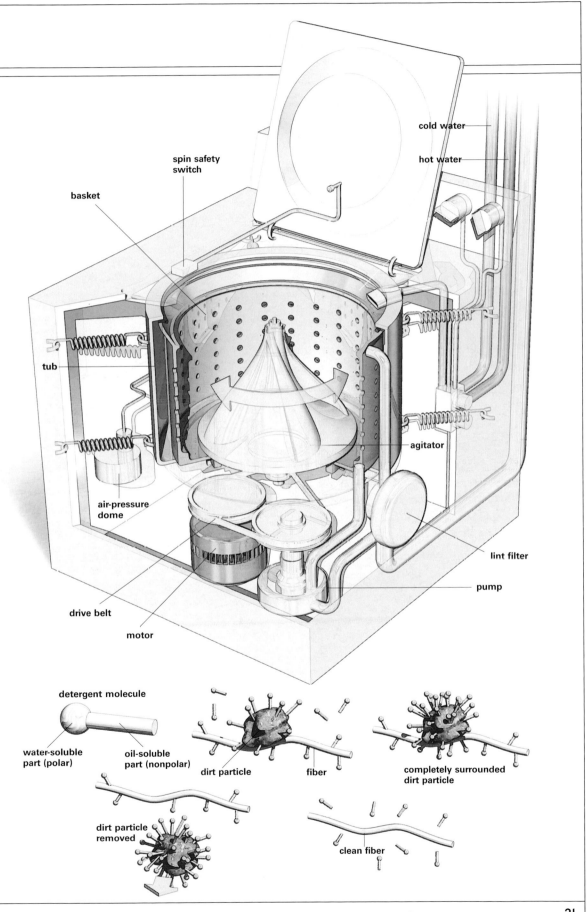

cold water

hot water

spin safety
switch

basket

tub

agitator

air-pressure
dome

lint filter

drive belt

pump

motor

detergent molecule

water-soluble
part (polar)

oil-soluble
part (nonpolar)

dirt particle

fiber

completely surrounded
dirt particle

dirt particle
removed

clean fiber

HOME HEATING & COOLING

I N TODAY'S BASIC HEATING SYSTEM, WHETHER GAS OR OIL, THE FUEL must be burned, and it must be mixed with air to do so. In a gas furnace, the gas-and-air mixture is ignited, and the heat produces hot water, steam, or hot air. A forced-air or hot-water circulating system spreads the heat through the home. Waste gases go up a flue. An oil burner also ignites a mixture of fuel and air; the fuel is first pumped from a tank and converted into a fine spray, then burned in a combustion chamber. Convector units then disperse the heat.

The simplest cooling device is the electric fan. It simply moves air around or pushes or pulls hot air out of a warm room. When fans are used with refrigerant, the coolants dissipate unwanted heat. Window units and central air conditioners remove heat and humidity and eject it outdoors, reducing the air temperature inside with the aid of a closed system containing a refrigerant. All such systems are based on these principles: When a liquid vaporizes, it draws heat from its surroundings; when a gas condenses back to a liquid under high pressure, heat is released.

In central air-conditioning, an evaporator coil is mounted on the furnace and a condensing unit rests on a concrete slab just outside the home. A small tube carries liquid coolant in from the condensing unit, and a larger tube returns the coolant in a gaseous state to the unit outside. The furnace blower draws in warm air from the house and sends it over the evaporator. As coolant in the evaporator changes from liquid to gas, it absorbs heat, thereby cooling the air, and causing it to lose moisture. The blower forces the cooled, dehumidified air into the supply ducts, and from there it circulates through the house.

CENTRAL AIR CONDITIONING
(Right, above) A concrete slab outside the house holds the condensing unit, which pumps cooled, liquid refrigerant through tubes connected to the evaporator in the house. Inside, a blower sends warm air from the home over the evaporator. As the coolant changes from a liquid to a gas, it absorbs heat and cools the air. A tube returns the coolant to the unit outside, while the blower forces the cooled air through supply ducts and into the home.

COOL DOWN
(Below) An Olympic hopeful at the 1996 trials in Athens, Georgia, cools off in front of a fan/mister in the summer heat.

SEE ALSO

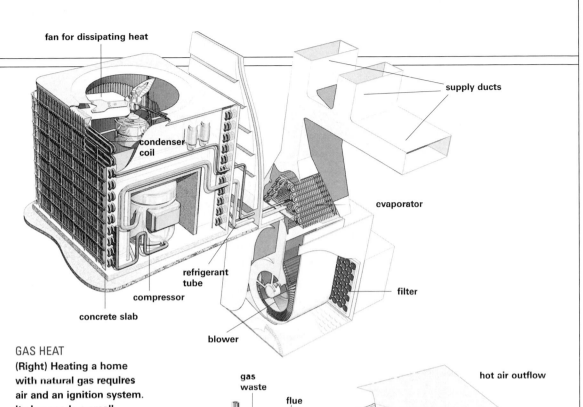

fan for dissipating heat

supply ducts

condenser coil

evaporator

refrigerant tube

filter

compressor

concrete slab

blower

GAS HEAT

(Right) Heating a home with natural gas requires air and an ignition system. It also needs a small, permanent flame called a pilot light. Gas flows into the system through a supply line, then to mixing tubes that combine it with air. From there, the mixture passes to burners, where the pilot light ignites it. An exchanger transfers the heat to water or air, while wastes escape through a flue. The thermocouple cuts off the gas if the pilot light goes out.

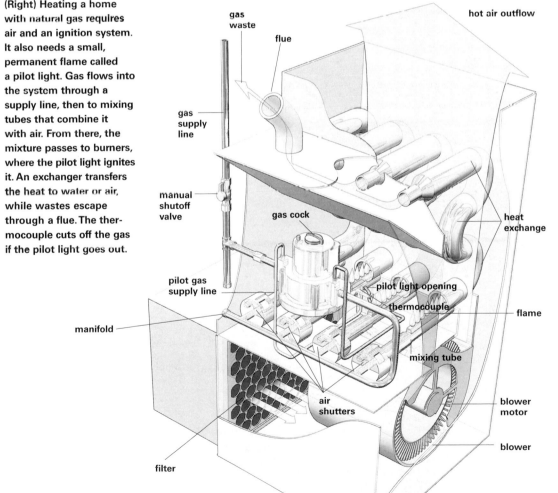

gas waste

flue

hot air outflow

gas supply line

manual shutoff valve

gas cock

heat exchange

pilot gas supply line

pilot light opening

thermocouple

flame

manifold

mixing tube

air shutters

blower motor

blower

filter

HAND TOOLS

SMALL HOUSEHOLD APPLIANCES AND POWER TOOLS REQUIRE FAIRLY strong motors that are capable of varying speeds. To fulfill its role, an appliance might have a unique combination of gears, cams, drive belts, pivots, shafts, rotors, and pinions. These connect the motor itself, directly or indirectly, to the various attachments that do things like dice an onion or turn a screwdriver.

Appliances often contain gears, which transmit motion and power from one part of a machine to another and may be called upon when more twisting and turning force than usual is needed. Reduction gears slow an appliance when too much speed is undesirable. Motors may also drive cams or other mechanisms that translate rotary into linear motion.

Many small appliances do not need gearing because their attachments work best when they draw power directly from the motor. While an electric mixer cannot be a direct extension of the motor's shaft because the speed would make a mess of the mix (a mixer employs a worm gear to slow down the action), a fan or the cutting blade in a food processor works directly off the motor, spinning at the same high speed.

When a homeowner turns a screw, he or she applies force to what is essentially a lever and, in so doing, produces torque, a rotational force. The effectiveness of the effort depends on the force exerted and the length of the lever's "arm." Thus, the handle of a screwdriver becomes more than a grip: It magnifies the force the hand uses to turn the blade and drive in a screw. The same thing is true of a pair of pliers twisting a nut. Power tools, such as drills and screwdrivers, draw extra strength from a nicely meshed system of gears attached to a chuck, a jawlike fixture that holds the drill bits and the screwdriver blades. Regulators control the flow of electricity to the motor's rotating armature.

LATHE

(Right) Chips fly as a carpenter shapes wood fixed to a lathe. This basic turning tool works by rotating an object, such as a piece of wood or metal, about a horizontal axis. A cutting tool moves across or parallel to the rotational direction, shaping the object as it turns.

ROUND AND ROUND

(Below left) Many household appliances rely on revolving wheels and gears. An electric shaver rotates a circle of blades on a springlike driveshaft that follows the contours of the skin. A garbage disposal unit grinds refuse on a turntable before flushing it through a drainpipe. An electric drill derives its power from a variable-speed, fan-cooled motor that turns its gears and the chuck holding the drill bit—or any of the tool's other attachments. The rotating commutator in the motor receives electricity through rubbing carbon brushes.

SEE ALSO

Power Stations · 32

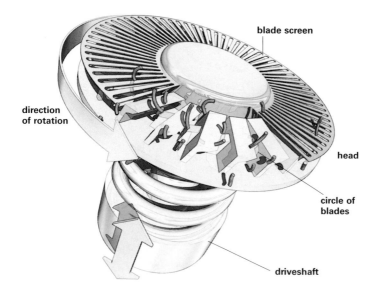

blade screen

direction of rotation

head

circle of blades

driveshaft

CLOCKS

HOW DID PEOPLE TELL TIME BEFORE THEY HAD CLOCKS? They observed stars at night, the shadow of a sundial's gnomon, marks made on a candle, or grains of sand sifting through an hourglass. The oldest mechanical clocks date from late-14th-century France.

Frills and variations aside, clocks are essentially boxes containing toothed wheels, pins, and springs, all precisely arranged to measure that impalpable continuum called time. In the basic clock, power is supplied by electrical impulses or by a falling weight or an unwinding spring, rewound when necessary. Movement of the hands is regulated by a train of toothed wheels, pinions, and spindles of different sizes, each taking its own appointed time to make a full revolution. In some clocks, a pendulum controls the wheel train's rate of rotation. In others, a mainspring turns a driving wheel that moves the wheels connected to hands on the face of the watch or clock.

In the quartz versions, a small electrical charge is applied to a quartz crystal; the crystal begins to vibrate and give off pulses of current in a precise, predictable manner. These pulses, in turn, can be used to control the motor turning a clock's hands or to advance the numerals displayed by the liquid crystals in a digital display. In a digital clock, alternating current powers the timepiece, and its precise cycles mark time's passage. Seven light-emitting diodes form numbers when energized in the correct combination.

Quartz vibrations are known as a piezo-electric reaction, and when they are put to work in a clock or a watch, they make for a timepiece that is far more accurate than a mechanical one. Couple the pulses of current from the oscillating quartz with a microchip that reduces their frequency to a usable rate—one pulse a second—and you have a clock or a watch that relies on minuscule electronic circuitry instead of toothed wheels and a mainspring.

While a quartz-crystal clock has an error rate of less than one-thousandth of a second per day, the latest generation of atomic clocks can be accurate to plus or minus one second in ten million years. Atomic clocks are faceless and handless wonders that trade on the constancy of the frequency of a molecular or atomic process. Put another way, atomic clocks use the extremely fast vibrations of molecules or atomic nuclei to take the measure of time. Because the vibrations remain constant, these clocks measure short intervals of time with much more precision than a mechanical clock.

CLOCKWORK
(Right) Powered by an unwinding mainspring, a clock consists of precisely machined gears and a hairspring that winds and unwinds to control the movement of the balance wheel, which regulates the escape wheel. A driving wheel turns as the mainspring unwinds, controlling the minute and hour wheels. The number of their teeth determines the number of revolutions that the hands will make.

PENDULUM
(Above) A pendulum's swing has a steady, measurable time span. Because successive swings occur in equal lengths of time, regardless of whether the swing is large or small, pendulums are ideally suited for controlling clock movements.

SEE ALSO

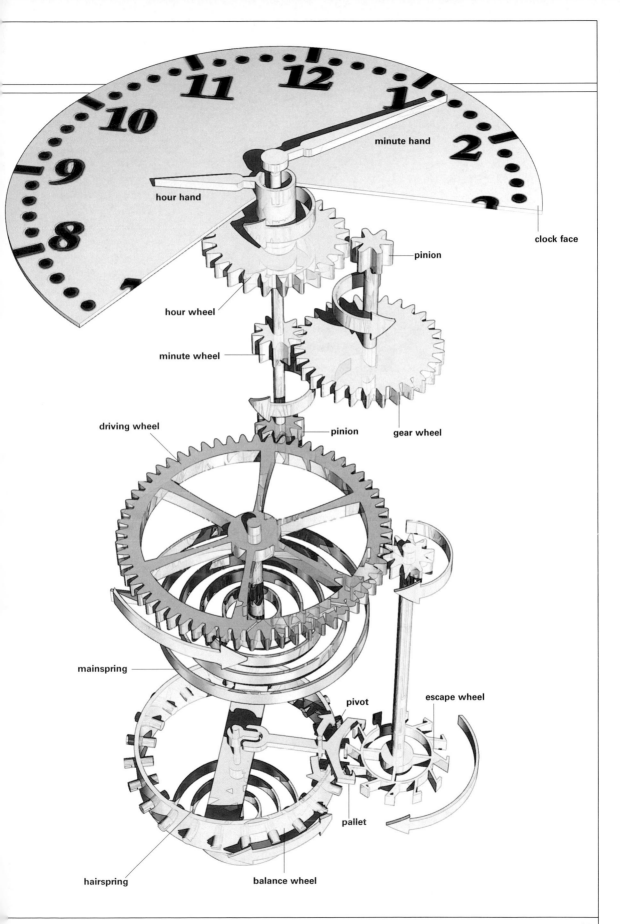

minute hand

hour hand

clock face

pinion

hour wheel

minute wheel

gear wheel

driving wheel

pinion

mainspring

pivot

escape wheel

pallet

hairspring

balance wheel

27

LIGHTBULBS

Before Thomas Edison's incandescent electric lamp, people depended on brushwood torches, candles, and oil lamps. Today, we brighten our homes with lightbulbs and fluorescent lamps and get around in the dark with flashlights, streetlights, and automobile headlights. Lights also help houseplants grow, readjust our internal body clocks after long flights, and soothe seasonal affective disorders.

Lighting has evolved in astonishing ways since Edison. In 1879, he passed electricity through a strand of carbonized cotton sewing thread, causing the rudimentary filament to glow for more than 13 hours in a glass vacuum tube. His feat was marred, though, by the loss of power through heat and by the short bulb life.

The coiled tungsten filament, a metal with a high melting point, made a brighter, longer-lasting bulb. The life of a bulb got even longer when an inactive gas—nitrogen, argon, or krypton—was placed inside to slow the filament's evaporation. A variation on the theme is the halogen lamp. Inside its bulb are molecules of bromine or iodine, halogen elements that combine with tungsten given off by the filament to form a gas. When this gas comes in contact with the hot filament, the tungsten atoms separate from the halogen and adhere to the filament, essentially rebuilding it.

But it is the energy-efficient, long-lasting LED (light-emitting diode) that now outshines other lightbulbs. LEDs are made of semiconductor material that produces light when subjected to an electric current. They are used in flashlights, electronics, traffic signals, Christmas tree lights, and more. LEDs have no filaments, which makes them burn cooler and last for many thousands of hours.

ELECTRIC LIGHTBULB
(Right) An incandescent bulb works because a body —in this case, a tungsten wire filament—gives off visible light when heated. Current passing through the filament heats the wire to more than 4000˚F, resulting in the radiation of electromagnetic energy.

BOTTLED LIGHT
(Below) Fluorescent light comes when electrons and mercury vapor atoms interact. Current to an electrode emits electrons that strike atoms of vapor and produces ultraviolet radiation. That bounces off the phosphor coating on the inside of the tube and energizes electrons in the phosphor's atoms. These atoms, in turn, radiate white light.

SEE ALSO
Power Stations · 32

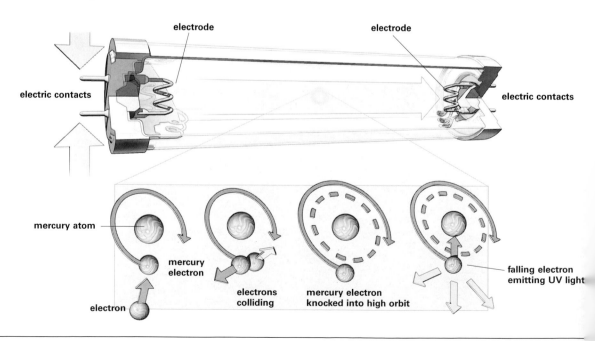

electrode   electrode

electric contacts   electric contacts

mercury atom

electron

mercury electron

electrons colliding

mercury electron knocked into high orbit

falling electron emitting UV light

CHAPTER 2

POWER & ENERGY

DEFINED AS THE CAPACITY FOR DOING WORK, energy takes many forms: It can be mechanical, electrical, physical, or thermal. When it is expended or dissipated by machines or human activity, we have power, which performs work. Sources of energy include water, wind, coal, oil, gas, wood, radioactive rocks, the moon's tidal pulls, and the sun's rays. Some people rely on biomass energy, employing waste-to-energy incinerators and other means to retrieve it. Coal and nuclear energy generate most of the electricity we use to light our buildings and power our machines. Water power supplies about a fifth of the world's electricity, while wind and geothermal energy are important sources of power for several countries. In the future, we may even look to exploding neutron stars for energy and, perhaps more immediately, to fusion—the same process that powers the sun.

Monuments to energy, cooling towers belch steam at a nuclear power plant in France.

POWER STATIONS

I
N 1886, WILLIAM STANLEY, A PIONEER IN THE GENERATION
and transmission of electric power, fired up what was probably
the earliest central power station. Erected in Great Barrington,
Massachusetts, his power station used a 25-horsepower boiler and
engine to turn a 500-volt generator and supply electric lighting to
25 businesses. Stanley's creation and other early power stations relied on
the same basic principles and components we use today, notably
fuel, boilers, turbines, and generators. They draw energy from a fuel
source, transform it to electricity, and send that power out into a grid.

Before the 1973 oil embargo, the United States depended on oil to
generate much of its electricity, but from 1973 to the 1990s, the use of
oil dropped from 17 percent to 3 percent. Coal is the fuel of choice at
most power stations today.

Coal arrives at most stations ground into dust, which is then mixed
with air to make a highly explosive fuel. Fired, it heats water in a boiler,
thus producing high-pressure steam that pushes against the propeller-
like blades of a giant turbine attached to the shaft of a huge generator.
As the shaft spins, coils of wire interact with a circular array of magnets
to create electricity.

Electricity then goes through a set of transformers, which are devices
that alter voltage. First, they step up the voltage for main distribution
along tall, latticed towers. Later, at a substation, before it is distributed,
other transformers bring it back down to the
240 and 120 volts normally used in homes
and offices.

POWER PLANTS
(Right) Power plants
help supply the world
with electricity, its most
versatile form of energy.
While plants can use
virtually any fuel to make
electricity, about 55
percent of the electricity
in the U.S. comes from
coal-burning facilities.

CURRENT
(Below) A simple AC
(alternating current)
generator consists of
magnets, a spinning
armature coil with
connecting slip rings, and
carbon conductor brushes.
It uses electromagnetic
induction to convert
mechanical energy into
electricity. As the coil spins
and the induced current
reverses in the magnetic
field, the current passes
from the rings to the
external circuit by way
of brushes that press
against the rings.

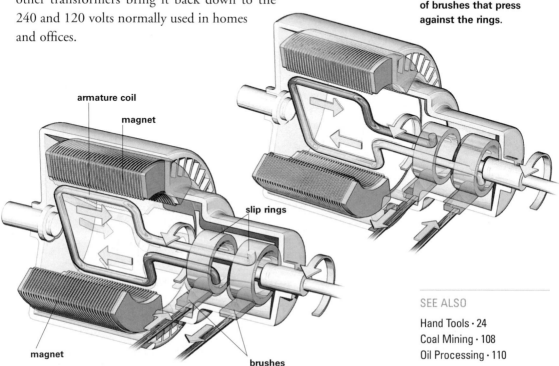

armature coil

magnet

slip rings

magnet

brushes

HYDROELECTRICITY

Place a waterwheel beneath a small waterfall. As it turns under the force of the water on its paddles, the energy can be shifted from its central, axle-like shaft to linked wheels, belts, and gears, and ultimately to useful tools. For centuries, people used water-power to grind corn and wheat, to spin fiber, and even to heat up their charcoal-burning, metal-smelting furnaces. The Chinese, for example, used water-powered bellows in A.D. 31.

A hydroelectric power station requires a large head of water and a fall or gradient to take advantage of gravity and the water's momentum. A hydroelectric plant is really a king-size version of the waterwheel, but its powerhouse contains generators and turbines whose blades spin under the carefully controlled force of a jet or flow of water.

Wave power, which has been used in Japan and Norway, takes advantage of the vast store of energy that is created in water when wind drives into it. Wave-to-energy devices called oscillating water columns (OWCs) do the work. Fixed in the water, when waves strike them, air inside is compressed and forced though turbines. Other versions float and drive pumps as they bob in the waves.

Harnessing the tides can be an expensive operation, but despite cost and their potential for damaging habitats, tidal power facilities are in use in France and a handful of other countries. The U.S. government funded a tidal project in the 1930s in Maine's Passamaquoddy Bay area. After dam gates trapped the incoming tide, the water would be released through turbine generators. Dikes were built, but politics and lack of funding put an end to the project.

THE FORCE OF WATER
(Right) Outlet tubes in the Glen Canyon Dam direct water from the Colorado River. Completed in 1966, the 710-foot-high dam supplies power and regulates the river's flow.

POWER STRUCTURE
(Below) Dams can produce electricity, store water for irrigation and public use, and control floodwater. This massive barrier controls water by regulating its flow through gates and within long tunnels and pipelines called penstocks or sluices. In a hydroelectric plant, water feeds through penstocks into the turbine. Bushings provide electrical insulation, and bus bars— lengths of conductors —collect and distribute the electric current.

SEE ALSO
Power Stations · 32

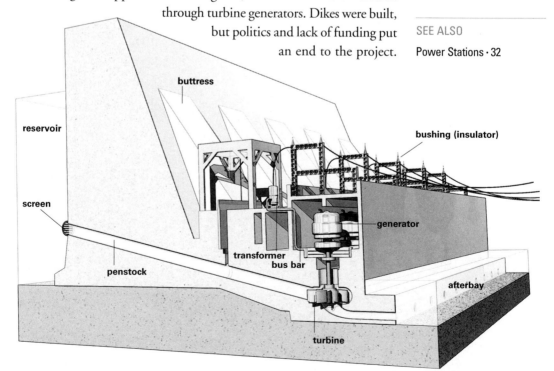

buttress

reservoir

bushing (insulator)

screen

generator

transformer
bus bar

penstock

afterbay

turbine

ALTERNATIVE POWER

WIND POWER CAPTURES THE MOVING AIR; GEOTHERMAL POWER taps steam and hot water underground; solar power gathers the sun's energy. Each requires special conditions, but where they are feasible, these alternative power systems work very well.

Like waterwheels, windmills have been used for centuries, their shafts connected to a series of gears and attachments that could mill grain, irrigate fields, and pump seawater from low-lying land. A wind turbine, the modern version of the windmill, drives a generator and produces electricity. The world's largest concentration of such turbines is in California.

Geothermal energy is the product of decaying radioactive elements in the ground. It can drive a turbine to produce electricity, or it can be piped into buildings for heating. In Iceland, many homes have underground heat piped in, but most underground heat is not so accessible or exploitable, and it is concentrated in reservoirs found only in certain parts of the world.

Plentiful, nonpolluting, and free, solar power is the energy emitted by the sun as electromagnetic radiation. Today, water or air heated by the sun fulfills a variety of household needs, providing hot water for kitchen sinks and showers or warm air to take the chill out of the house itself.

Capturing solar energy to heat a home can be done passively or actively. Passive heating works best in a well-insulated home with large, south-facing windows (in the Northern Hemisphere), an interior of dark stone or tile, and a location in a sunny region. Such a home receives lots of sunlight, which warms the air inside; the dark interior absorbs the light and reradiates it as heat during the night.

Active solar heating requires equipment that collects, stores, and distributes solar energy. A flat-plate collector fixed to the roof of a house serves either an air- or a water-based system. A clear cover, usually glass, allows sunlight to pass through and strike a dark, heat-absorbing metal plate. As the plate soaks up heat, it warms air or water in pipes or tubes running through the collector. In an air-based system, a fan blows the warmed air into a rock-filled storage area under the house. A water-based system pumps the warm water into a storage tank for household use or through tubes inside floors or ceilings for space heating.

SUN WORSHIP

(Right) Arrays of PV (photovoltaic) cells collect the power of the sun and convert it to electricity for household and industrial uses. The photovoltaic process occurs inside small cells constructed of silicon and boron. When sunlight strikes these materials, it causes a flow of electrons, which is captured and turned into useful electrical power. Solar energy plants now provide power for more than a quarter of a million American homes.

SEE ALSO

Home Heating & Cooling · 22
Power Stations · 32
Alternative Fuels · 58

EARTH'S OWN ENERGY

A tempting energy source, geothermal power cannot warm everyone. Adequately powerful sources are few and far between. Often the sources manifest themselves quite dramatically as geysers, steam vents, fumaroles, and hot springs, valued therapeutically for centuries. But surface displays may be far from a reservoir, and some sources of geothermal energy give no sign of their presence, requiring prospectors to use seismic and geologic probes that are complex and costly. Only certain regions hold natural reservoirs of very hot water, and even those sources will run dry, may contain contaminants, and would cost a great deal to develop.

NUCLEAR POWER

(Left) The cooling stacks of a nuclear plant spew nonradioactive steam, not smoke. Steam spins the electricity generator, and changes back to water to cool the system.

REACTOR RODS

(Right, above) Twelve-foot-long rods initiate and control a chain reaction. Fuel rods contain uranium in ceramic pellets; the rods themselves are made of zirconium, a metal that resists heat, radiation, and corrosion. Control rods, shown in red, absorb neutrons, preventing them from hitting and splitting uranium atoms. Inserted, these rods slow the chain reaction; withdrawn, they speed it up.

A CHAIN REACTION

(Far right, above) The process begins when neutrons smash into uranium atoms in the fuel pellets. The atoms fission, or split, releasing neutrons of their own. One fission spawns other fissions, which trigger more. As the atoms split, they liberate intense heat. In an atomic bomb, the reaction moves quickly, creating an explosion. In a reactor, the speed of the reaction is carefully controlled.

SEE ALSO

A SINGLE OUNCE OF THE URANIUM-235 ISOTOPE CAN GENERATE an immense amount of energy, a fact that makes it easy to understand why more than 400 nuclear plants in 30 countries—a quarter of them in the United States—rely on it to produce electricity. Nuclear energy now provides about 20 percent of the electricity in the United States, and the distinctive domes and towers of nuclear plants have become sights as familiar as gas stations and shopping malls.

Each plant generates electricity with the same steam-turbine-generator arrangement used by a coal-fired plant. The fuel, however, consists of solid ceramic pellets containing isotopes of uranium atoms that are split apart in a process known as fission.

The pellets are packed in long metal tubes, fuel rods, which are bundled together and installed in a heavily shielded and water-cooled reactor, where fission occurs. When uncharged subatomic particles called neutrons are released into the reactor, the particles collide with the uranium atoms and split them. The nuclei of the uranium atoms burst, and the atoms release their own neutrons, which then strike other atoms, and on and on, creating a chain reaction. The immense heat generated by the process vaporizes water, and the steam that is created turns the turbine and generator, arranged as in a coal-fired plant to produce electricity.

The operation of a nuclear power plant is controlled by the use of rods made of neutron-absorbing material. By inserting and removing these rods, plant operators slow down or speed up the reaction going on inside the nuclear reactor. Inserting the rods prevents neutrons from hitting other atoms, and thus slows down the nuclear reaction, while withdrawing the rods allows the process to speed up.

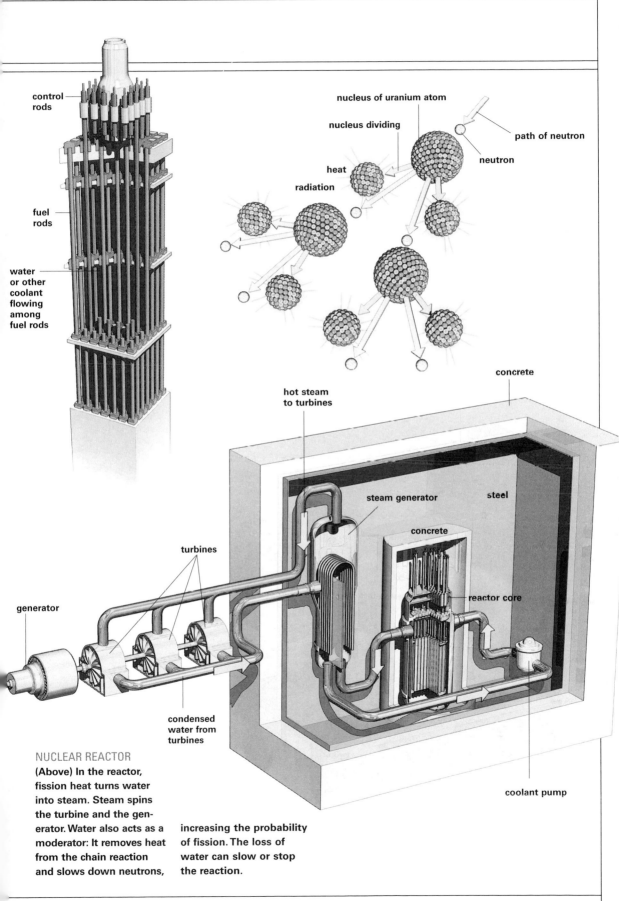

control rods

fuel rods

water or other coolant flowing among fuel rods

nucleus of uranium atom

nucleus dividing

path of neutron

neutron

heat

radiation

concrete

hot steam to turbines

steam generator

steel

concrete

turbines

reactor core

generator

condensed water from turbines

coolant pump

NUCLEAR REACTOR
(Above) In the reactor, fission heat turns water into steam. Steam spins the turbine and the generator. Water also acts as a moderator: It removes heat from the chain reaction and slows down neutrons, increasing the probability of fission. The loss of water can slow or stop the reaction.

FUSION

FUSION IS THE NUCLEAR REACTION THAT ENERGIZES NOT ONLY THE sun and stars but also the hydrogen bomb. It occurs when the nuclei of small, light atoms are compressed under intense heat to form larger and heavier nuclei, creating enormous bursts of energy.

The fuel to start the fusion process is plentiful: Deuterium and tritium, the heavy isotopes of hydrogen, can be extracted from ordinary seawater. So in theory, this energy could be captured and converted into electricity, a process that appears simple on paper. But the problem is how to control such titanic energy.

Reactors would be required to withstand heat from plasma, the seething gas of charged particles resulting from a fusion reaction. Scientists have tested experimental containers made of coils of wire—"magnetic bottles" that create magnetic fields powerful enough to confine the reaction. The doughnut-shaped Tokamak Fusion Test Reactor at the U.S. Department of Energy's Princeton Plasma Physics Laboratory—shut down in 1997 after 15 years of operation—had 587 tons of coils of superconducting materials in a riblike arrangement. At 24 feet tall, with a diameter of 38 feet and an 80-ton vacuum chamber, it generated a horizontal field that forced charged plasma particles to circle inside the container. The particles collided, fused, and released energy without touching the chamber walls. The reactor set fusion records, reaching 510 million degrees Celsius, more than 25 times the temperature at the sun's center.

ITER, a newly designed tokamak fusion reactor, is now being built as an international venture in Cadarache, France.

INSIDE THE TOKAMAK
(Right) Princeton University researchers attempt to achieve the elusive break-even point at which energy released by fusion equals that required to produce it. Building plasma containers has proved a daunting task, given the furious heat they must hold. But with superconducting materials that contain heat more efficiently than copper, fusion power may live up to expectations as a source of safe and limitless energy.

TOKAMAK
(Below left) In 1951 Soviet physicists Andrei Sakharov and Igor Tamm proposed the tokamak, a device to contain hot plasma, fusion's charged gas. Inside the tokamak (short for "toroidal chamber with an axial magnetic field"), superconducting materials create a magnetic field that causes plasma to flow within the doughnut-shaped chamber.

SEE ALSO

central solenoid magnet

protective shielding

access port

plasma

access port

toroidal field magnet

poloidal field magnet

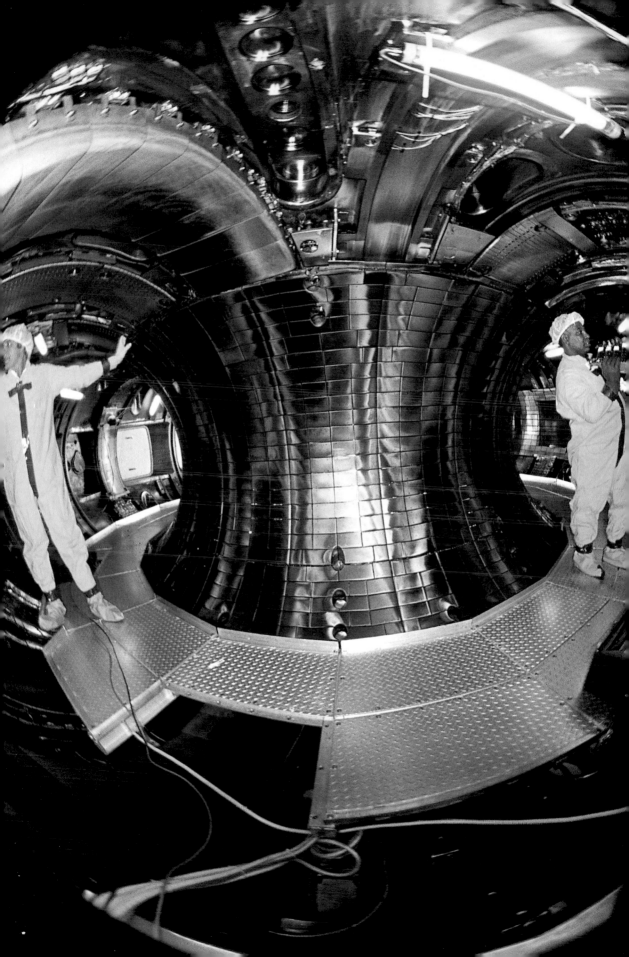

CHAPTER 3

BUILDING

TODAY'S CONSTRUCTION PROJECTS OFTEN USE wood, brick, stone, and concrete—the same materials ancient builders relied upon. To erect buildings of great size, complexity, and beauty, many projects also rely on iron, steel, aluminum, glass, and synthetic materials that were never dreamed of by early architects. Simplicity of design and function remain, but building today is an exact science, different from the days when construction was a practical craft based on experience and observation alone. Modern buildings, along with older ones that have been modified, are engineered to withstand winds, earthquakes, traffic, hurricanes, and fire, and most are probably far more comfortable, safe, and efficient than anything that was built in the past. The equipment used to erect these structures is also impressive. Occasionally, it is even more dazzling than the buildings it is shaping.

Stark against the sky, an ironworker, girders, and a crane typify a building site.

SKYSCRAPERS

A N AMERICAN ORIGINAL, THE SKYSCRAPER BEGAN ITS gravity- and wind-defying rise in the late 19th century, a soaring testimonial to economic boom times, mass production, and the new technology of structural engineering.

Developed first in Chicago, Carl Sandburg's "city of the big shoulders," the skyscraper relied on a steel skeleton of columns and beams rather than masonry walls to support its weight. A "curtain wall" of nonbearing materials, such as glass and thin marble sheets, sheathed the framework, resulting in relatively lightweight buildings with excellent tensile strength. Innovations such as I-shaped steel beams, concrete reinforced with embedded steel, and tubular concrete designs allowing lighter and stronger walls enabled skyscrapers to rise even higher.

Once, the main things that concerned a builder were slope, drainage, and proximity to a water supply. Site preparation amounted to clearing away trees, underbrush, and rocks. Excavation, if any, was minimal, and a foundation was laid rather quickly using wood, bricks, mortar, and hand tools.

Today's builders take more into consideration: Faults can trigger earthquakes; soil can heave or sink structures when changes in weather or soil composition occur; high water tables can flood cellars; radon may be present in rocks. Tools are now nail guns, power saws, routers, jackhammers, and electronic surveying devices. On larger-scale construction projects, they are dynamite, water pumps, and earth-moving equipment. A variety of cranes can work on the ground or atop skyscrapers.

To protect against the lateral forces generated by earthquakes or winds, engineers today construct nonrigid buildings that, in effect, sway with the forces. One technique places layers of rubber and steel between base and foundation. Another relies on bearings slid under load-bearing columns, which help dissipate a quake's energy through friction. Tubular concrete design actually transforms a building's exterior walls into rigid tubes that effectively carry gravity and wind loads. Special skyscraper windows are designed and built to prevent shattering and separation from the framing.

But even with new techniques and materials, skyscrapers still must stand on solid bedrock, well-compacted soil, and, within them, deep concrete foundations to bear the building's enormous weight. Engineers must consider the "dead load"—the total weight of the structure and of all its fixed equipment—and the "live load," which changes all the time and includes vibrations caused by equipment, people, even moving furniture along with seismic and wind forces.

Whatever the technology used to build a skyscraper, the fact remains that erecting one can take years of proper planning.

ON THE JOB
(Right) Building a high rise requires the vision of architects, adequate urban land, tons of material, and a king's ransom. It also demands the skills of steelworkers, stonemasons, glaziers, welders, electricians, plumbers, and countless other workers seemingly oblivious to doing their jobs at great heights, often exposed to the elements.

SEE ALSO

ELEMENTS OF CONSTRUCTION

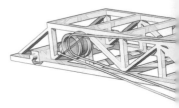

T HE KEYSTONE IS A SIMPLE, WEDGE-SHAPED BLOCK OF STONE ON which much depends. Fitted last into the top of an arch, it locks the other pieces in place, providing support for the curved structure that, in turn, supports weight above it.

Keystones and all the other basic elements of construction work as a team to ensure stability and distribute weight. The result can last for centuries: Among the arch's exemplars of form and function are the aqueducts built by the ancient Romans. They curved across valleys and cities and carried water to baths and fountains from sources far away.

The forces at work in an arch, or in any building—bending, compression, tension, and torsion, or twisting—create internal stresses. Structural forms abound, each with its own function, to deal with these stresses. Columns support beams and lintels; horizontal spans carry the load of a roof. Piers, arches, rib-vaulting, and buttresses, as well as flying buttresses and semidomes, absorb thrust and support huge domes.

Most structures adhere to principles that preserve structural integrity, as well as to stringent building codes that enforce them. Homes and buildings referred to as modular are also governed by the elements of construction. Contrary to what some believe, modular homes are not "mobile" homes, which may be built on a steel frame with wheels. Modular buildings—which include homes, hospitals, correctional facilities, military housing, churches, and restaurants—are true to their name. Their structural components, called modules, are manufactured in a controlled factory environment, and then assembled with the help of a crane into a complete structure on a site foundation. Constructed of the same materials that go into build-on-site "stick homes" (the term for a conventionally built dwelling), a modular building can be finished in a few weeks after it is set on the foundation.

A CLIMBING TOWER CRANE

(Right) This crane builds itself section by section, in effect hauling itself up by its own bootstraps as the building it constructs gradually rises. A counterweight balances the horizontal jib so that the tower, anchored in a concrete base, does not tip over when the crane lifts a load.

CONSTRUCTION SITE

(Left) The site at first gives little hint of the splendid edifices that will rise above it. But even before the buildings take shape, work sites invariably attract passers-by to the noisy drama in the below-ground arena.

SEE ALSO

Hand Tools · 24
Skyscrapers · 44

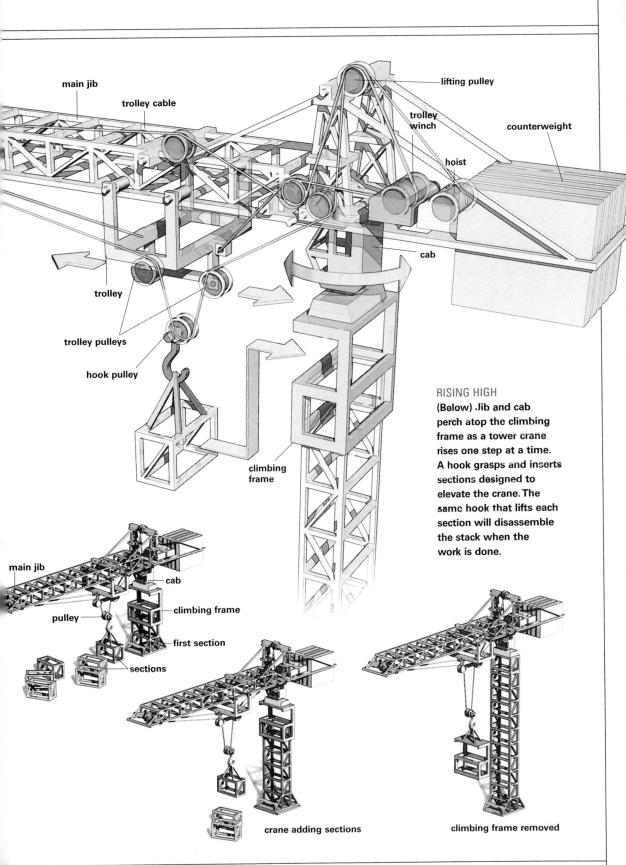

main jib

trolley cable

lifting pulley

trolley winch

counterweight

hoist

cab

trolley

trolley pulleys

hook pulley

climbing frame

RISING HIGH

(Below) Jib and cab perch atop the climbing frame as a tower crane rises one step at a time. A hook grasps and inserts sections designed to elevate the crane. The same hook that lifts each section will disassemble the stack when the work is done.

main jib

cab

pulley

climbing frame

first section

sections

crane adding sections

climbing frame removed

47

ELEVATORS & ESCALATORS

PERHAPS THE MOST ESSENTIAL TECHNOLOGY IN A HIGH-RISE building is the people mover. Elevators and escalators make skyscrapers possible. They have had an extraordinary economic impact on businesses and public facilities, and they are factors to be reckoned with when architects design new buildings.

The first elevator was developed in the middle of the 19th century by Elisha Otis, an American inventor. Powered by steam, his device ran up and down between guide rails and had an automatic safety device that clamped on the rails to prevent the car from falling if the hoisting rope broke. Hydraulic systems powered later elevators. Some small buildings still have them, but most models now operate with steel hoist cables drawn by an electric motor over a grooved pulley wheel, called a sheave. The cables attach to a counterweight that goes up when the car descends.

Escalators, on the other hand, are moving stairs mounted on a continuous chain that is drawn over a drive wheel run by an electric motor. The stairs form a sort of level treadmill at the beginning and end of the ascent, becoming an arrangement of treads and risers during the incline. While elevators are swifter and can fit into small wells inside a building or run up its sides, escalators do not stop and can carry more people.

This conveyor-belt technology—the same that has moved coal, sand, and grain for years—is also used in another people mover: the moving sidewalk, commonly found in airports today. An early model built by a coal company was unveiled at the 1893 World's Columbian Exposition in Chicago and used to transport visitors past exhibits. While moving walkways are common fixtures, some airports also use automated driverless vehicles that run over steel or concrete guideways. Powered by electricity and guided by computers, the cars stop at the right stations at the right times and the doors open and shut automatically.

PATHS CROSS
(Right) Shoppers enjoy the ease of an escalator network in a department store in Osaka, Japan.

INCLINED ROUTE
(Below) Stairs and handrails appear and disappear as an escalator moves effortlessly over its inclined path. Built on the endless chain principle, an escalator relies on an electric motor at the top. A chain—drawn around a drive wheel at the top of the landing and a return wheel at the bottom—moves the treadmill-like stairs attached to it. As the stairs move, they form treads and risers. They flatten out at the end of the stair span and again at the beginning.

SEE ALSO
Skyscrapers · 44

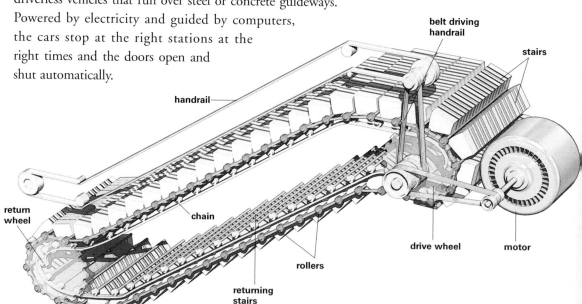

belt driving handrail

stairs

handrail

return wheel

chain

drive wheel

motor

rollers

returning stairs

TUNNELS

DIGGING A TUNNEL INVOLVES A SLOW, SEEMINGLY BLIND, PASSAGE of equipment and workers underground—through a mountain, under streets, or beneath an expanse of water—with the crews and machines excavating, blasting, and reinforcing their way from the two opposite ends until they meet.

Tunneling requires heavy cutting equipment, such as huge tunnel-boring machines (TBMs) that have rotating drill heads. Moved slowly along by gigantic hydraulic rams and guided by computer-linked lasers and many other electronic distance-measuring contrivances, these engineering marvels simultaneously cut through and eject tons of soil and rock. They also line the tunnel with prefabricated concrete or cast-iron segments to keep it from caving in.

For at least two and a half centuries, people in England and France dreamed of digging a tunnel under the English Channel—and more than a few thought that the dream was a nightmare. To them, a tunnel linking the two shores would serve more than peaceful pursuits. It might also provide an avenue of invasion. But in the most expensive privately financed engineering project in history, TBMs completed the 31-mile-long, three-tunnel "Chunnel" linking the two countries in 1994. The tunnel's 24-mile underwater section, burrowed into a chalk layer well beneath the Channel, is the longest in the world. More than just a tunnel, it has two-way rail tubes for shuttles, freight trains, and high-speed passenger trains. The Chunnel's central service tunnel provides fresh air and allows for emergency evacuation and maintenance.

Designed to bore smooth walls to an exact size, TBMs eliminate blasting accidents and deafening noises. They are very expensive, though. An alternative is to lay precast tunnel sections into trenches where they are joined and backfilled. Despite the high-tech safety precautions of either method, tunneling remains one of the most dangerous engineering projects that can be attempted.

TUNNEL-BORING MACHINE
(Right) One of the 11 mammoth boring machines on the Channel Tunnel project dwarfs workers. Each machine's rotating drill head cut through rock 200 feet below the floor of the English Channel.

CHANNEL TUNNEL
(Below) Embedded in a chalk layer, the "Chunnel" contains two one-way rail tunnels for trains and shuttles and a central service tunnel that provides ventilation, room for maintenance, and an evacuation route. Crossover caverns allow trains to change tracks, while piston relief ducts balance pressure waves caused by fast-moving traffic. Cross passages carry air from the service tunnel to main tunnels.

SEE ALSO
Electric & High-Speed Trains · 64

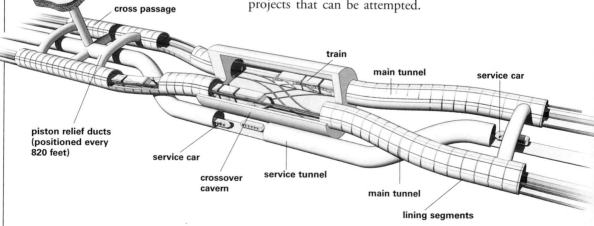

ventilation

cross passage

train

main tunnel

service car

piston relief ducts (positioned every 820 feet)

service car

crossover cavern

service tunnel

main tunnel

lining segments

BRIDGES

WHETHER THE QUAINT, WOODEN, COVERED STRUCTURES that still span some of our rivers and streams, or magnificent stretches of steel and heavy cable that loom high over deep bays and valleys, all bridges have the same function: to provide easy passage over natural and artificial obstacles.

There are five general types of bridges: beam, arch, suspension, cantilever, and cable-stayed. Which type of bridge to build depends on many factors—span, winds, vulnerability to earthquakes and tides, required clearances, and the weight of traffic. Building bridges requires a keen understanding of loads and stresses, metal fatigue, strength of materials, and soil and sediment science. The choice of materials and style of the bridge are always vital considerations.

A beam bridge (a span supported at either end, like a log across a ditch) is the most common form of bridge design. When resting merely on supports at either end, its length is limited due to the bending effect of a downward load on the beam. So for longer spans, additional supports, or piers, and spans are added. The arch bridge puts more stress on the end supports than does an ordinary beam, but it can accommodate longer spans. Since its shape makes it virtually impossible for ordinary traffic to pass over it, a flat deck usually hangs above or below it.

Suspension bridges, because of the efficient way they handle and distribute loads, have no equal in the distance they can cover. These bridges have several key parts: towers set on piers, steel cables that arc between them, steel hangers that connect the traffic-carrying deck to the structure and carry load up to the cables, and steel and concrete anchorages that hold the cables at either end. The decks' loads are transferred from the cables to the towers, and weight is sent to the bridge's foundations. These bridges are not always practical because they can be expensive to build.

Cantilever bridges are stiffer than suspension bridges and do not have attachment sites at opposite ends. Instead, two rigid, deep beams extend from opposite piers, like brackets. Fixed at one end only, they join in the middle to form one weight-bearing surface. The cable-stayed bridge is characterized by diagonal steel cables, called stays, which attach the bridge's deck to tall, mastlike towers that sit atop weight-absorbing foundation piers.

THE GOLDEN GATE
(Right) Inching his way along a high cable on San Francisco's Golden Gate Bridge, a worker heads up through a fog bank to do some painting.

SEE ALSO

Elements of Construction · 46
Steelmaking · 106

BRIDGE DESIGN

Engineers must take into account both the bridge's own weight and the weight of the traffic it will bear in order to calculate the total requirements of the foundation. They choose among the various designs to build a structure best suited for the stresses and strains posed by the setting and the predicted use of the bridge. Builders know that when any force is imposed on a structure, it tends to deform. They likewise know that materials respond in different ways when a load is applied. Thus style and materials are vital considerations in any bridge design. Further, because a bridge is in a sense a landmark and a monument, aesthetic form plays a part in bridge design as well.

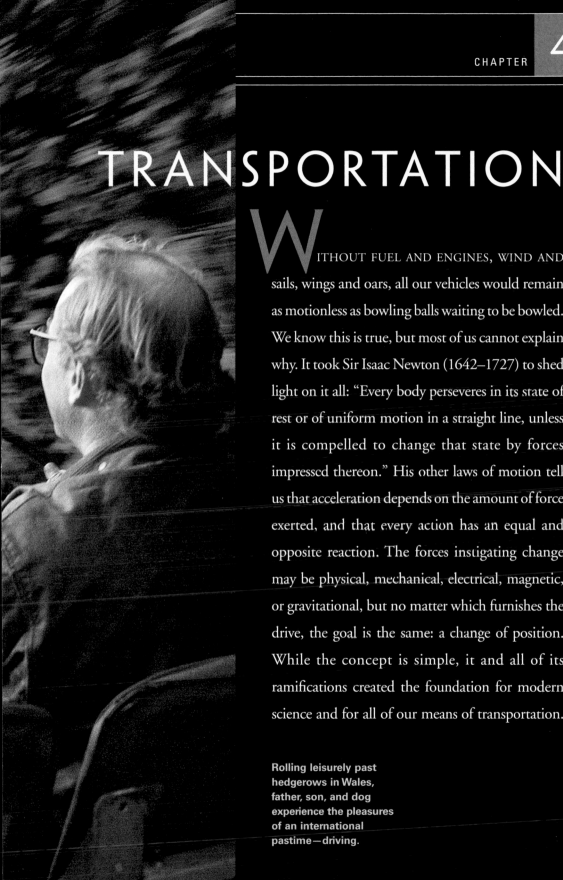

CHAPTER 4

TRANSPORTATION

WITHOUT FUEL AND ENGINES, WIND AND sails, wings and oars, all our vehicles would remain as motionless as bowling balls waiting to be bowled. We know this is true, but most of us cannot explain why. It took Sir Isaac Newton (1642–1727) to shed light on it all: "Every body perseveres in its state of rest or of uniform motion in a straight line, unless it is compelled to change that state by forces impressed thereon." His other laws of motion tell us that acceleration depends on the amount of force exerted, and that every action has an equal and opposite reaction. The forces instigating change may be physical, mechanical, electrical, magnetic, or gravitational, but no matter which furnishes the drive, the goal is the same: a change of position. While the concept is simple, it and all of its ramifications created the foundation for modern science and for all of our means of transportation.

Rolling leisurely past hedgerows in Wales, father, son, and dog experience the pleasures of an international pastime—driving.

AUTOMOBILES

THE MODERN AUTOMOBILE IS A FAR CRY FROM LEONARDO da Vinci's 15th-century concept of a steam-propelled vehicle, and a long way from the three-wheeled, three-mile-an-hour steam carriage built in 1769 by Nicolas-Joseph Cugnot, a French Army officer. Early cars, some electrically driven, were produced in small numbers just before the start of the 20th century. Many of the models were made in Europe, and only a wealthy few owned them. The Duryea Motor Wagon Company, founded in 1895 in Springfield, Massachusetts, built 13 cars that first year. Today, many millions of passenger cars are in operation in the United States, a testimonial not only to the overwhelming demand for comfortable, reliable, and economical transportation, but also to mass production and a revolutionary manufacturing process: the assembly line.

Two Americans share credit for supplying the demand. Beginning in 1901, Ransom Eli Olds made the United States' first commercially successful car, the Oldsmobile. But the car that captured the popular imagination was Henry Ford's lightweight yet strongly built Model T. First manufactured in 1908, it was the first automobile built by modern mass-production methods. Engine and chassis assemblies were tested on one level of the factory and driven to the bottom of a chute, where the auto bodies were slid down and bolted onto the frame. Then the cars were driven off the assembly line and out to eager buyers.

Today's modern automobile assembly line is a round-the-clock combination of human assemblers and computerized operations that may reach ten miles in length. Along the way, giant presses shape and stamp out parts, and robots weld the automobile frame. The body is dipped into a rustproofing bath, and coats of paint are baked on.

Many industrial robots are dangerous to work with, and assemblers don't always relish being too close to them. One new type of assembly-line robot, developed in response to this problem, is called a cobot, short for "collaborative robot." Developed by researchers at Northwestern University, a cobot provides only guidance while workers provide the power. "Intelligent assist devices," as their engineering creators call them, cobots help workers guide heavy and unwieldy parts—doors, windshields, and seats—into position without damaging the car's body. Cobots provide computer-directed "virtual surfaces," invisible boundaries that guard, say, a doorframe and the interior of the car, redirecting an auto part if it gets pushed in the wrong direction by a worker.

CARBURETOR AND ENGINE
(Right) An automobile's carburetor teams up with the engine (below right) to ignite and burn vaporized gasolines. The carburetor delivers the correct fuel and air mixture into the cylinders, where a spark ignites it. Small explosions move the pistons, the first steps toward moving the car.

AUTOWORKERS
(Above) Unhindered by doors and windows that special machines will add later, workmen install essential parts and systems in the suspended space frame of a nearly completed car.

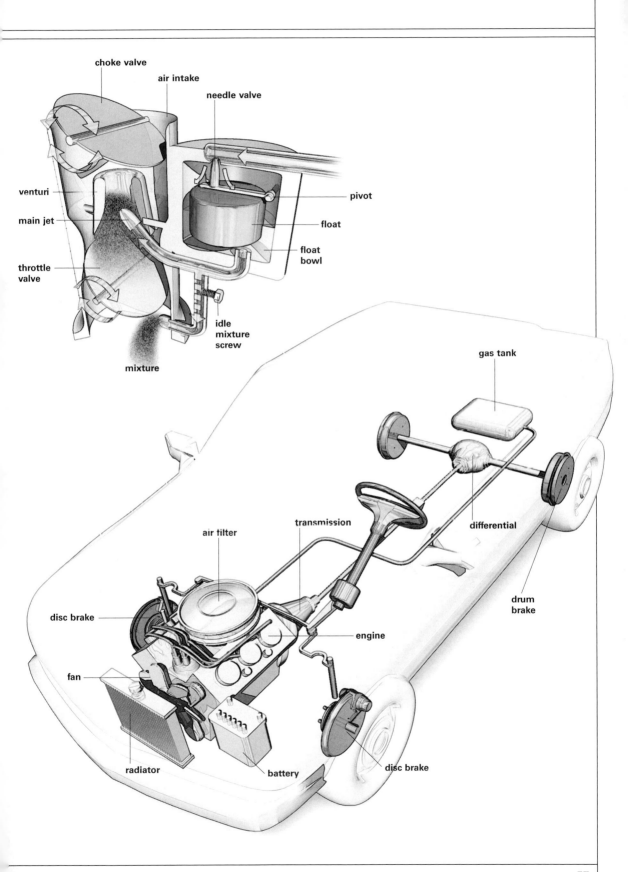

choke valve

air intake

needle valve

venturi

main jet

throttle
valve

pivot

float

float
bowl

idle
mixture
screw

mixture

gas tank

air filter

transmission

differential

disc brake

drum
brake

engine

fan

radiator

battery

disc brake

ALTERNATIVE FUELS

TWO OF TODAY'S MAJOR CONCERNS ARE AIR POLLUTION, CAUSED by gasoline-powered automobiles, and our limited fossil fuel supply. Engineers are looking to electricity for help on both fronts. The ordinary gasoline-powered car already relies on electricity—up to a point. Its electrical system furnishes electricity to operate the starter, ignition system, and accessories, and it recharges the storage batteries. But a gasoline-powered automobile needs pistons and a carburetor, a water pump and a muffler, none of which are needed in an electric car.

A simpler machine, the electric car sends electrical energy stored in the battery directly to the motor, where it is converted into mechanical energy. An onboard recharger plugs into an outside electrical outlet for "refueling." Owners of electric cars do not have to deal with oil changes, but their batteries generally take two or three hours to recharge after a hundred or so miles of road time. With improved batteries, or possibly batteries that store solar energy, these cars may one day become more popular.

Forward-looking drivers now operate hybrid automobiles that use fuel cells or electric motors in combination with gasoline engines. A fuel cell is basically a pair of electrodes wrapped around an electrolyte. It generates heat and electricity by combining oxygen and hydrogen without combustion. The hydrogen fuel powering the cell is actually separated from the car's gasoline, but the fuel cell utilizes so much more of the gasoline's potential energy than a piston engine does that its range may be doubled on the same amount of gas, with less pollution. Even more environment-friendly is hydrogen removed from water by sunlight, a feat already accomplished by scientists.

Some hybrid cars use the gasoline engine to charge the batteries that run the electric motor. Others switch between engine and motor, according to driving conditions—starting on battery power, for example, and running on a combination of gasoline and electric power when they speed up.

The automotive industry won't be giving up gasoline and diesel fuel in the near future, given the enormous effort of revamping refineries, gas stations, parts manufacturers, and more. Yet technology and science are never-ending quests—and scientists of the future are certain to make the gasoline engine a relic, like the horse and buggy it replaced.

THE ELECTRIC CAR
(Right) As the technology advances, electric cars will become as elegant as their gas-powered counterparts. With few moving parts, their engines are remarkably quiet.

SEE ALSO

AUTO ELECTRONICS
More than 80 percent of automotive innovation is now based on increasingly complex electronics. In the modern car, so-called data networks—which enable the engine to communicate with other parts of the vehicle—rely on over 80 microprocessors and even more "power" semiconductors to direct all the mechanical and safety systems, such as airbag and antilock brakes, steering, and antitheft devices. Perhaps the most valuable system is the OBD, which stands for on-board diagnostics. OBD uses computers to monitor the performance of major engine parts, alerts drivers of problems with dashboard messages, and provides mechanics with scan tools to diagnose internal problems.

BICYCLES

RIDING A BICYCLE IS ONE OF THE SIMPLEST WAYS TO TRAVEL: It generally requires a minimum of expertise beyond knowing how to shift gears and keep one's balance. But the bicycle, which dates back more than 200 years, is an invention whose relatively humble appearance belies its complexity.

The early bike had iron tires that were first propelled by the rider's feet, then by ropes wound around an axle and hitched to a lever for driving power. The seat was uncomfortable, to say the least, because it was usually no more than a wooden beam. Even with refinements, the first bikes were frequently ridiculed: One 18th-century model was dubbed a "dandy horse" by Englishmen who thought the frame resembled a pack animal. A modern bicycle, however, can be a high-tech, designer dream-machine equipped with a score of gears, an aerodynamic configuration, airfoil tubing, and a titanium cantilever frame.

In a sense, the rider is a bicycle's engine, supplying the physical energy to be converted to mechanical energy. The bike's chain serves to transmit force from one place to another, in this case from the pedal sprocket to the rear driving wheel. Gears on the front sprocket and the rear wheel serve the same purpose as other gears: They change one rate of rotation to another. This makes the bicycle very efficient, because the effort applied to the pedals can be geared up for high speed or geared down for hill-climbing power. A pair of devices called derailleurs transfers the chain from one sprocket to another. In highest gear, the rear wheel turns many times for each turn of the pedals, and the bike moves along swiftly; in low gear, when a stronger forward shove is required to go, say, up a hill, the wheel turns fewer times in relation to the pedal, thereby trading speed for ease of pedaling.

THE HUMAN BODY
(Below) Muscle power is all that a bike needs to get moving, although new materials and designs help make bikes more aero-dynamic and cycling easier. Frames may be constructed of aluminum, lightweight metal alloys, or carbon fiber composites, with shock absorbers of various polymers.

SEE ALSO

Hand Tools · 24
Automobiles · 56

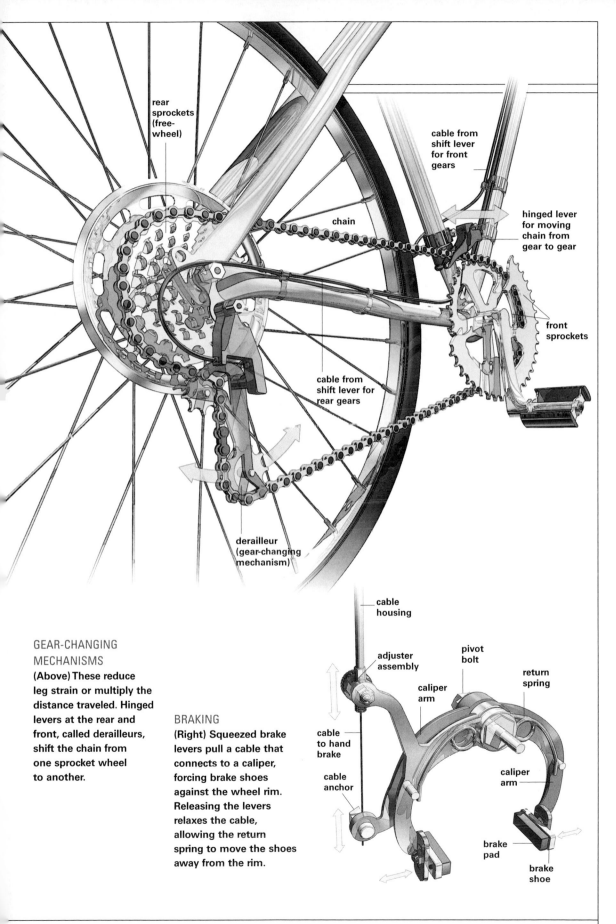

rear
sprockets
(free-
wheel)

cable from
shift lever
for front
gears

chain

hinged lever
for moving
chain from
gear to gear

front
sprockets

cable from
shift lever for
rear gears

derailleur
(gear-changing
mechanism)

cable
housing

GEAR-CHANGING
MECHANISMS
(Above) These reduce
leg strain or multiply the
distance traveled. Hinged
levels at the rear and
front, called derailleurs,
shift the chain from
one sprocket wheel
to another.

adjuster
assembly

pivot
bolt

caliper
arm

return
spring

BRAKING
(Right) Squeezed brake
levers pull a cable that
connects to a caliper,
forcing brake shoes
against the wheel rim.
Releasing the levers
relaxes the cable,
allowing the return
spring to move the shoes
away from the rim.

cable
to hand
brake

cable
anchor

caliper
arm

brake
pad

brake
shoe

TRAINS

T RAINS ARE WONDERFUL," THE MYSTERY WRITER AGATHA CHRISTIE once observed. "To travel by train is to see nature and human beings, towns and churches and rivers, in fact, to see life." A 19th-century steam locomotive was a coal stoker's backbreaking nightmare, though, and wasted energy, dirtied the air with soot, and cost a lot to operate. One such juggernaut ran at a mile a minute in the early 1900s, burned three tons of coal before pulling out of the roundhouse, carried four tons on a tender, and required 4,000 gallons of water in its tank.

With its piston rods connected to driving wheels as much as 85 inches in diameter—an arrangement akin to a tricycle's pedal wheel—the steam locomotive is the perfect visible example of how a machine can convert one form of movement into another. As the driving rods move back and forth under the force of steam, their linear movement is shifted into rotary mode, turning the wheels.

The diesel engine spelled the end of the line for the steam locomotive. Put on the market in 1898 by inventor Rudolf Diesel, a German mechanical engineer, it relies on the same piston strokes and basic moving parts as the gasoline engine but burns a heavier, thicker, less expensive fuel and does not need spark plugs for ignition.

Because a diesel engine performs more work per gallon of fuel, it is a boon for long-hauling trucks, buses, ships, and trains, and for heavy agricultural and road-building equipment. In a road vehicle the engine's power is transmitted directly to the wheels, but in a train the engine is connected to an electrical generator. The current produced is stored in huge batteries, then fed to electric motors installed in so-called bogies, which are pivoting carriages that not only house the motors for the driving wheels but also enable the train to negotiate curves.

RIDING THE RAILS
(Above) The Amtrak Cascades races between Seattle and Portland. Diesel trains eliminate concerns about overhead electric wires exposed to weather and steam engine smoke.

LOCOMOTIVES
(Right, above) Steam locomotives burn fuel in a firebox. Heat passes through tubes inside the boiler and generates steam, which enters the cylinders, moves the pistons, and drives the train. In diesel locomotives (right, below) fuel and electricity provide power without the huff and puff of steam. The engine connects to a generator, making electricity for storage in large batteries alongside the wheels.

SEE ALSO
Electric & High-Speed Trains · 64

BOILER WORKS

(Below, top) Essentially a furnace and boiler on wheels, a steam locomotive burns fuel, usually coal, in a chamber called a firebox. Heat passes through tubes inside the huge water-filled boiler and generates steam, which is collected and sent through U-shaped "superheater" tubes to cylinders on each side of the locomotive's front end. As high-pressure steam enters the cylinders, it moves the pistons and drives the train.

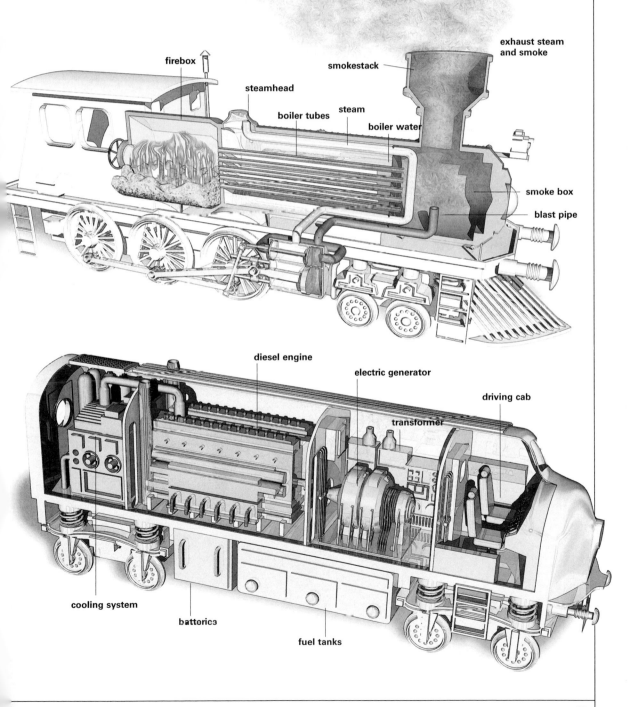

firebox

steamhead

boiler tubes

steam

smokestack

boiler water

exhaust steam and smoke

smoke box

blast pipe

diesel engine

electric generator

driving cab

transformer

cooling system

batteries

fuel tanks

ELECTRIC & HIGH-SPEED TRAINS

I N 1873, ANDREW SMITH HALLIDIE, AN AMERICAN ENGINEER AND inventor, patented the first cable cars: rickety trolleys hauled up San Francisco's steep hills by an endless wire-rope cable running in a slot between the rails. Drawn by a steam-driven mechanism in a powerhouse, the cable cars—still operating today but with technical improvements— eventually spelled the end of horsepower as a means of moving passenger coaches. Other cities, including Los Angeles, Washington, D.C., and Kansas City, followed suit, but traffic congestion was a major problem.

The electrified subway changed all that. The world's first subway, run with steam locomotives, opened in England in 1863. It was converted to electricity in 1893. That year Budapest, Hungary, became the first European city to build an electric subway line. America's first electric subway system opened in Boston in 1897, followed in 1904 by the one in New York City.

Subway trains run on a pair of rails, generally powered by voltage running through a third rail. Electric surface trains work on the same principle, but they usually pick up voltage from overhead lines with a folding, scaffoldlike apparatus known as a pantograph. The same arrangement drives electric buses, sometimes known as trackless trolleys. Monorails, elevated trains built for relatively short distances, run on a single rail, either straddling it or hanging beneath it.

Top speed for a subway is around 75 miles an hour, although in 1893 a steam engine thundered through New York State at 112.3 mph. Today Amtrak's Acela can travel at about 150 mph, and France's relatively lightweight TGV (Train à Grande Vitesse) can hit 185. Automatic speed enforcement systems supervise the drivers of these superfast trains.

UNDERGROUND

(Below) An underground Mass Transportation Rail train rolls into a station in Hong Kong. Essential to public transport, electric railway systems also have miniature—and playful— counterparts in remote-controlled, low-voltage systems. Owners move their cars about with small transformers that increase and reduce voltage.

SEE ALSO

Power Stations · 32
Trains · 62

SUPERCONDUCTIVITY

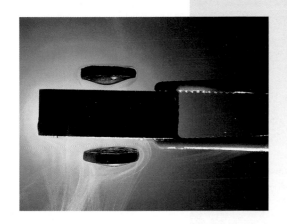

(Left) A thallium-based superconductor bathed in super-cold nitrogen vapor floats magnets above and below it, thanks to repellent magnetic force. A superconducting metal transports electricity indefinitely without loss from resistance when subjected to intense cold, and it produces a magnetic field.

In 1979, Japanese scientists set a speed record of 361 miles an hour with a maglev train that had superconducting magnets on board. Superconductors will eventually affect all uses of electricity. As this technology goes into routine use, trains, computers, and other machines will run faster and faster.

THE MAGLEV TRAIN

(Below) Electromagnetic levitation —maglev for short—provides the most speed over the rails. Electromagnets on the train's underside and on the track float the vehicle on a magnetic cushion with no ground contact. Fed with alternating current, propulsion magnets in a U-shaped aluminum guideway pull and push the train along and control its speed. Swift, comfortable maglev trains give new meaning to "ground transportation."

train magnets

track magnets

train magnets

track magnets

track

train magnets

HOW BOATS STAY AFLOAT

CCORDING TO LEGEND, THE GREEK MATHEMATICIAN AND physicist Archimedes (287-212 B.C.) was sitting in his bathtub, taking note of the amount of water his body had caused to overflow onto the floor, when he formulated the basic principle of buoyancy. Thus began the science of hydrostatics, which deals with the laws of fluids at rest and under pressure. Its main point is that a body immersed in fluid loses weight equal to the weight of the fluid displaced. Put another way, things that float—ships and swimmers, ducks and dugouts—do so when the weight of water they displace is exactly the same as their own. If, however, an object displaces a weight of water less than its own, it sinks.

But there's more to it than weight. A wooden or fiberglass platform will not float if too many heavy rocks are put on it, and neither will a bracelet or chunk of steel placed on the water's surface. They sink because their density is greater than that of water and because they do not displace enough water to create the upward force necessary for flotation. On the other hand, wooden boats and steel ships float because their hollow configuration makes their average density less than that of water. Their weight is spread over a larger volume and they displace more water than, say, a steel block. We float, too, when we inhale and rest on our backs in a lake, and we sink when we exhale and point our feet downward.

BUOYANCY

(Above) Oblivious to the rules of hydrostatics, floaters enjoy the moment. They do what they do because of the ability of water, or other fluid, to thrust upward when a body is placed on it. Saltwater has greater buoyancy than fresh because it is heavier.

SEE ALSO

Sailboats · 68
Submersibles · 70

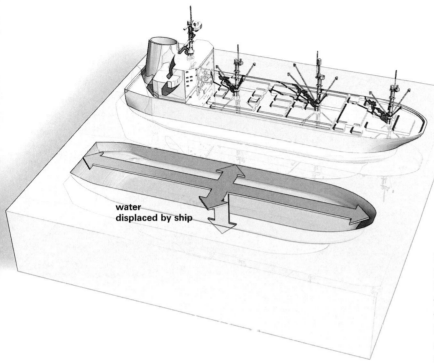

water
displaced by ship

DISPLACEMENT

(Left) A ship floats because
it creates an upthrust force
from the water equaling
the vessel's own weight.
For example, a thick, flat
slab of steel manufactured
without a hollowed-out,
bowl-shaped hull to
distribute the slab's weight
would quickly sink like
a stone. Design can also
affect speed; the deeper
a boat sits in water, the
slower it goes.

STABILITY

(Below) The steadiness
of a ship depends on the
vessel's center of gravity.
Keels and pontoons help
maintain stability and
minimize the risk of

capsizing, but very large
vessels may employ
stabilizers, winglike
protuberances that
provide lift to counteract
the sea's roll.

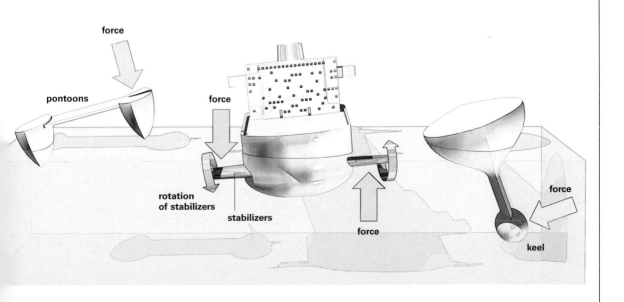

force

pontoons

force

rotation
of stabilizers

stabilizers

force

force

keel

SAILBOATS

WITHOUT THE WIND, BECALMED SAILBOATS ARE NEARLY AS STILL as seascapes hanging preserved in time on museum walls. Paddles or a motor could get things going, but only the sail conjures up images of graceful movement, nature's own technology at work. "Of all man-made things," wrote a sailor-author, "there is nothing so lovely as a sailboat."

Whether a sailing vessel is a simple day-sailer or an elaborate, full-rigged merchantman or man-o'-war carrying hundreds of yards of canvas, it is a wind-driven work of art. Its sails act as airfoils: It is a nautical version of an airplane. Lift, or drive, results when wind exerts less pressure on the convex side of a sail. A boat sailing into the wind is pulled diagonally forward by lift generated as wind flows over the sails; a boat running before the wind—that is, moving in a direction such that the wind blows from behind—is shoved forward by wind pressure.

The complex array of sails and rigging used by the classic square-riggers—20 or more enormous sails in a three-masted ship—provides more drive for the mainsails by compressing, funneling, and deflecting air. The auxiliary sails may also furnish drive themselves. Sailing is still largely a hand-controlled endeavor, but winches may be motorized, and automated and hydraulic systems can hoist, reef, and furl sails, all of which enable one person to handle a sailboat of almost any size.

It is the sail itself, however, that must do the lion's share of the work. Today, Dacron is the cloth of choice for most sails; templates are used to achieve the sail's right size and shape (called the belly) from triangular to quadrilateral. Sewn with heavy waxed thread, the sails have metallic eyes set into the corners through which rope can be passed. Sail tape reinforces the sail's edge, which often meets the wind head-on.

CATCH THE BREEZE
(Right) Sailboats moving in the same direction as the wind can gain even more power by raising a spinnaker, a lightweight, billowy sail that balloons out above the boat's bow. By presenting so much more surface area to the wind, spinnakers can increase boat speed significantly.

MANEUVER WITH WIND
(Below) Sailing before the wind is an easy push forward; sailing forward against the wind, or with wind abeam—blowing across—requires complex maneuvers, such as tacking: setting a zigzag course and changing the sail's angle with every turn.

SEE ALSO
How Boats Stay Afloat · 66
Principles of Flight · 72

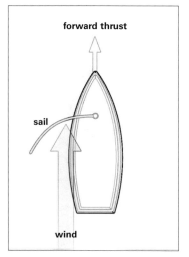

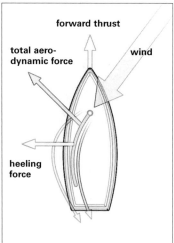

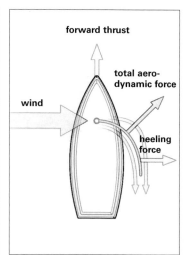

SUBMERSIBLES

Navy submarines float, dive, and rise by adjusting the amount of water and air in their ballast tanks. On the surface, with tanks full of air, a ship weighs less than the volume of water it displaces. Flooding the tanks causes a sub to sink, because it comes to weigh more than the water it displaces. A vessel rises by forcing air into its tanks and expelling water. To float just beneath the surface, the water in the tanks must equal the weight of the water displaced.

While the technology behind nuclear propulsion is highly classified information, one can safely say that a submarine travels essentially on steam. Intense heat is produced from the fission of nuclear fuel in the ship's heavily shielded reactor. This heat generates steam, which drives the turbine generators that supply electricity to the ship and the main propulsion turbines that turn the propeller. The generators pull fresh water and oxygen from seawater, so crews can live submerged for months.

Research submersibles that do bathymetry (bottom topography) and other ocean chores may be manned or remotely operated vessels (ROVs) tethered to a research vessel. Built to withstand the crushing pressures in the deep sea, ROVs are very maneuverable and carry a wide array of tools for observation, retrieval, collection, and repair.

One unique ocean expedition device is the drifter, a waterproof tube for data collection and transmission. Arms extending from the tube have vinyl or cloth sails stretched between them; floats suspend the sails beneath the surface, where they catch a current. When the drifter is freed, its transmitter signals to a polar satellite that relays its data to a receiving station. Sensors aboard the drifter can also measure surface temperatures, ocean color, pressure, and salinity.

MINISUB

(Right) Passengers aboard a miniaturized submarine near Grand Cayman Island in the Caribbean can travel to depths of a thousand feet below the ocean's surface. Minisubs support research as well as hunts for submerged shipwrecks and lost treasure.

SUBMARINE DESIGN

(Below) A rounded configuration and double-walled hull resist crushing deepwater ocean pressures. A submarine can dive, travel under-water, rise, and run on the surface—all by adjusting the volume of water and air in ballast tanks. A nuclear reactor generates steam that drives turbines and turns the propeller.

SEE ALSO

Nucear Power · 38
How Boats Stay Afloat · 66
Satellite Communications · 156

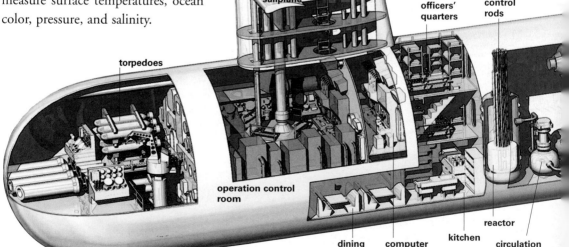

periscopes

antennas

conning tower

sailplane

officers' quarters

control rods

torpedoes

operation control room

dining room

computer room

kitchen

reactor

circulation pump

SONAR

(Below) The acronym, which stands for "sound navigation and ranging," describes a device or a method that transmits electronically generated sound waves through water. Computers tune to the pinging echoes, or reflected sound waves, to determine the direction from which they come and calculate the time they take to return. That information can be used to navigate, measure depth, map the seabed, and detect underwater objects.

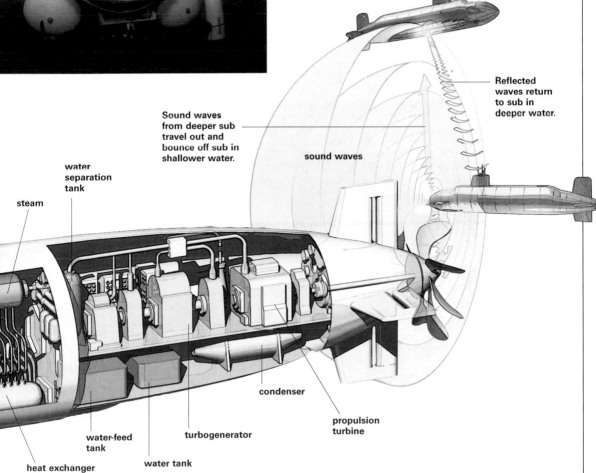

Reflected waves return to sub in deeper water.

Sound waves from deeper sub travel out and bounce off sub in shallower water.

sound waves

steam

water separation tank

condenser

propulsion turbine

turbogenerator

water-feed tank

water tank

heat exchanger

PRINCIPLES OF FLIGHT

EVER SINCE THE ANCIENT CHINESE BEGAN TO EXPERIMENT WITH man-flying kites and parachutes, and perhaps even earlier, people have tried to emulate birds. Indeed, the same aerodynamic principles and forces govern birds and aircraft: lift, drag, thrust, and weight.

Air hitting a wing's leading edge and streaming over top and bottom surfaces of the airfoils, or wings, provides the lift that raises a craft and keeps it "afloat." Enter Bernoulli's principle: Pressure is inversely related to velocity. Put another way, fast-moving air exerts less pressure than slow-moving air. During flight, the airflow over the longer upper surface of a wing travels faster than air on the underside, producing less pressure. The net force on the wing is upward, exerted by the slower-moving stream of air underneath.

Lift keeps the craft aloft. Drag, a force caused by friction as the plane moves through the air, slows the plane and requires the thrust of an engine to compensate and keep the plane moving. Providing thrust is expensive, and, to study ways to do so economically, engineers use wind tunnels to study the lift, drag, control, and stability of various designs.

An engineless glider is a somewhat different breed of aircraft. Generally towed aloft by a plane and cut loose to glide, it requires only drag, lift, and weight. Gliders are designed to descend slowly from a higher altitude to a lower one. Glider operators search for pockets of air (updrafts) that rise faster than the glider is descending, thus increasing potential energy. Ground heat creates one kind of updraft: thermals, in which hawks, too, circle and gain altitude without moving their wings.

AIRLIFT

(Right) Air that flows over the top of a wing moves faster and exerts less pressure than the stream beneath; the greater pressure of the slower-moving air below the wing lifts the airplane.

LIFT

(Below) A jetliner heads for the sky using forces different from those that keep ships afloat. Planes must generate lift greater than their own weight during takeoff—and equal to their own weight to stay aloft. Speed and wing surfaces provide the great lifting force needed to take off.

SEE ALSO

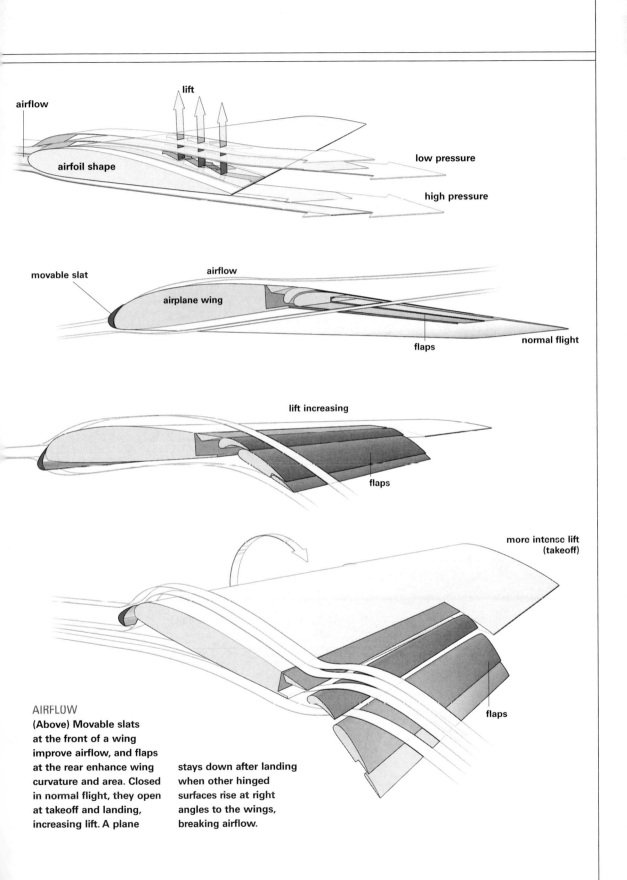

lift

airflow

airfoil shape

low pressure

high pressure

movable slat

airflow

airplane wing

flaps

normal flight

lift increasing

flaps

more intense lift
(takeoff)

flaps

AIRFLOW
(Above) Movable slats
at the front of a wing
improve airflow, and flaps
at the rear enhance wing
curvature and area. Closed
in normal flight, they open
at takeoff and landing,
increasing lift. A plane
stays down after landing
when other hinged
surfaces rise at right
angles to the wings,
breaking airflow.

PROPELLERS & JET ENGINES

T HE WRIGHT BROTHERS' PIONEERING FLYING MACHINE WAS powered by a 4-cylinder, 12-horsepower gasoline engine; its carburetor was a tomato can. Airplanes that came along later were able to get by without tomato cans, but they also relied on simple internal combustion engines to drive propellers that converted engine shaft torque, or turning force, into thrust. Propellers have blades that are shaped like wings: The front surface of each blade is more curved than the back, so a forward aerodynamic force is produced.

A jet engine works in much the same way, although it doesn't look like a piston engine. A commonly used example explaining some of the principles behind its operation is adequate only insofar as it goes: If you inflate a balloon and release it untied, it will fizz about a room until the escaping air is depleted. A jet engine is basically an internal combustion engine, but it uses the energy produced by combustion directly, and it does not need pistons to transmit driving power. Large quantities of air are drawn into the engine, compressed by a bladed turbine, and sprayed with kerosene fuel. When the mixture is ignited and the temperature in the combustion chamber exceeds 2500°F, the heated, expanding gases rush through an exhaust nozzle, providing the tremendous thrust that drives the plane forward.

AERODYNAMIC RELIC
(Above) A single-engine biplane, double-winged for extra support, skims over a cloud bank. Used during World War I as fighters, biplanes with thin wings were effective, but they were soon replaced by monoplanes with thicker wings.

SEE ALSO

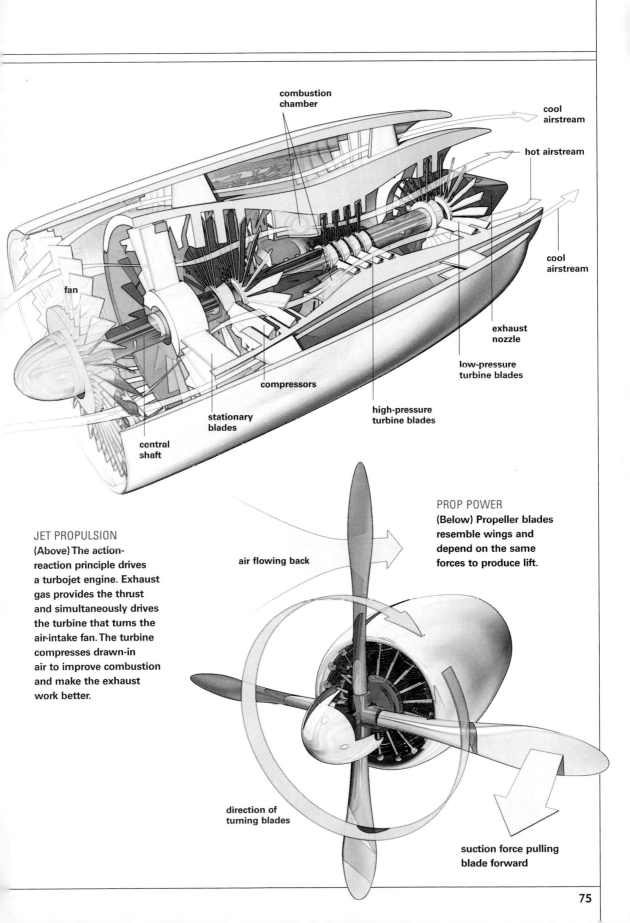

combustion
chamber

cool
airstream

hot airstream

cool
airstream

fan

exhaust
nozzle

low-pressure
turbine blades

compressors

high-pressure
turbine blades

stationary
blades

central
shaft

PROP POWER

(Below) Propeller blades
resemble wings and
depend on the same
forces to produce lift.

JET PROPULSION

(Above) The action-
reaction principle drives
a turbojet engine. Exhaust
gas provides the thrust
and simultaneously drives
the turbine that turns the
air-intake fan. The turbine
compresses drawn-in
air to improve combustion
and make the exhaust
work better.

air flowing back

direction of
turning blades

suction force pulling
blade forward

HELICOPTERS

LEONARDO DA VINCI IS CREDITED WITH DESIGNING A FLYING machine that could twirl itself skyward with the aid of a helix, or screw. He was inspired, perhaps, by "bamboo dragonflies"—the name given by fourth-century Chinese to toplike toys that could climb into the air with the pull of a string. Da Vinci could not build a working model, though, because the engine to power it had not been invented.

The first helicopter, invented by Igor Sikorsky, a Russian-born U.S. aircraft designer, flew in 1939. The basic design was pure Leonardo, and the principles behind it were the same that govern all aircraft. The difference is in the helicopter's rotor blades, by which it can ascend and descend vertically, hover, and fly in any direction.

Because a helicopter's moving wings provide both lift and thrust, the angle at which each blade enters the air, called the pitch, is essential for control. As the blades turn, the pilot uses the collective pitch stick to increase their pitch equally and thus lift the helicopter vertically; when the pitch decreases, the helicopter descends. To hover, the blades are angled just enough to produce lift equal to the craft's weight.

Propelling a craft forward, backward, or sideways requires use of a cyclic control stick that tilts each blade precisely. Such ease of operation has a downside: the action-reaction law. With only an overhead rotor, a helicopter would be forced in the direction opposite the rotating blades and would spin out of control. To compensate, a tail rotor shoves air to one side and keeps the craft on the right path.

Helicopters are workhorses for search and rescue, hoisting, traffic reporting, firefighting, and police work. Guided by global positioning satellites, robot copters run with computerized sensing and vision systems.

WHIRLYBIRD WORKHORSE
(Below) A Chinook helicopter lifts a load of ammunition for the Army's First Cavalry Division. Versatile military and civilian craft, helicopters can spot traffic, serve as weapon platforms, rescue survivors, evacuate ill patients, and hoist enormous loads.

SEE ALSO
Principles of Flight · 72
Propellers & Jet Engines · 74

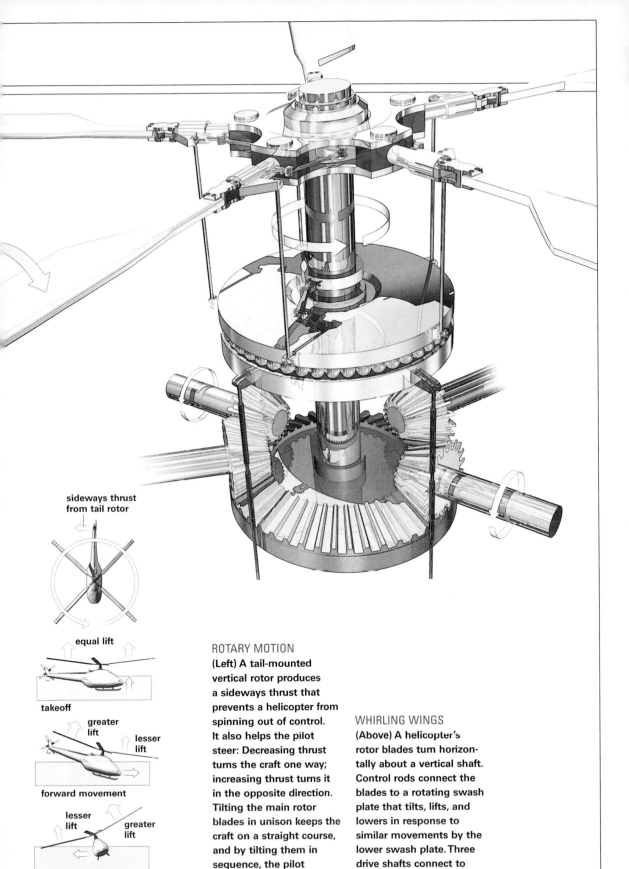

sideways thrust
from tail rotor

equal lift

takeoff

greater
lift

lesser
lift

forward movement

lesser
lift

greater
lift

flying sideways
to change direction

ROTARY MOTION

(Left) A tail-mounted
vertical rotor produces
a sideways thrust that
prevents a helicopter from
spinning out of control.
It also helps the pilot
steer: Decreasing thrust
turns the craft one way;
increasing thrust turns it
in the opposite direction.
Tilting the main rotor
blades in unison keeps the
craft on a straight course,
and by tilting them in
sequence, the pilot
increases lift on one side
to move left or right.

WHIRLING WINGS

(Above) A helicopter's
rotor blades turn horizon-
tally about a vertical shaft.
Control rods connect the
blades to a rotating swash
plate that tilts, lifts, and
lowers in response to
similar movements by the
lower swash plate. Three
drive shafts connect to
engines and drive the
main shaft via bevel gears.

SUPER- & HYPERSONIC PLANES

SOUND TRAVELS AT 1,090 FEET A SECOND IN AIR, A FIGURE THAT translates into Mach 1 in an aeronautical engineer's lexicon. The first time an aircraft flew faster than that was in 1947, when Capt. Chuck Yeager of the U.S. Air Force took the controls of an experimental rocket-powered Bell X-1. Today, while military fighters routinely break through the sound barrier, only one commercial airliner has flown at supersonic speed: the hundred-seat, fuel-guzzling, ear-splitting Concorde, which streaked through the skies at Mach 2. The Concorde ended service in 2003, the victim of rising maintenance and fare costs.

Experimental hypersonic planes can now fly at Mach 7 to Mach 10. The X-43, a futuristic-looking craft fueled by hydrogen, is propelled by an air-breathing engine known as a scramjet. A conventional rocket engine burns hydrogen and oxygen; the scramjet carries no oxygen, nor do fan blades compress air for forward movement, as in a normal jet engine. The scramjet sucks in atmospheric air for combustion.

The space shuttle is another high- and fast-flying hydrogen-oxygen burner. First flown in 1981, the winged shuttle is a multiengine, rocket-boosted, reusable craft that relies on aerodynamics and the laws of physics to speed at 17,500 mph in orbit around Earth.

The shuttle is equipped with two solid rocket boosters (SRBs) that each carry a million pounds of solid propellant materials. At launch the SRBs attach to a huge external tank loaded with more than a half-million gallons of super-cold liquid oxygen and liquid hydrogen. These are drawn into the shuttle's three main engines at the spacecraft's aft end, then mixed and burned in a combustion chamber. With thrust from the SRBs (which provide most of the force) and lift by the engines, the entire enormous bundle of fuel, guidance and life-support systems, computers, and crew streaks to an altitude of between 190 and 330 miles above sea level. A few minutes after launch, the SRBs detach and parachute into the ocean while the main engines continue to fire. Minutes later, the engines shut down, the external fuel tank separates, and a pair of orbital maneuvering system (OMS) engines fire. When the shuttle returns to Earth, its OMS engines shift its position and slow it down for a high-speed reentry. The commander takes over from the computer controls, drops the landing gear, and makes an unpowered descent and landing, just as a glider would.

LIFTOFF
(Right) The space shuttle *Columbia* launches from the Kennedy Space Center in Florida. When it returns, the shuttle will make an unpowered landing, like a glider.

FUTURISTIC FLIGHT
(Below) The *Horus,* a German-designed spacecraft, was designed to go from a conventional runway takeoff into orbit on its own rockets. Cost prohibited *Horus* from ever flying, but such rocket ships may ply space in future decades.

SEE ALSO
Alternative Fuels · 58
Principles of Flight · 72
Propellers & Jet Engines · 74

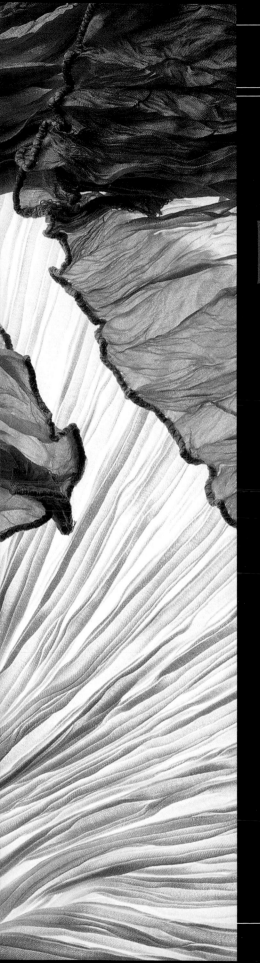

CHAPTER 5

MATERIALS

FOR AS LONG AS THERE HAVE BEEN TOOLS, there have been materials—the very stuff those tools were designed to work. Humans first gleaned raw materials from nature—rock, wood, plant fibers, wool from sheep, silk from worms. They learned to harvest materials from hidden realms of the Earth—metals, minerals, gems, coal, and, in the modern age, gas and oil. Artisans developed ways to turn natural substances into remarkably plastic materials, like glass and clay—and from those to shape everything from commonplace objects to immensely valuable treasures. Today we drown in a sea of material, especially plastic, the consummate material of our petroleum-dependent age. Researchers of the future will design new materials, to be sure, but they will also seek smart new ways to transform and reuse materials already at hand.

Fine pleats in sheer silk characterized the early 20th-century creations of Mariano Fortuny.

SYNTHETIC FIBERS

SYNTHETIC FIBERS WERE FIRST MADE IN THE LATE 19TH CENTURY, when chemists learned that cellulose could be extracted from wood pulp and formed into thread through nozzles. Originally called artificial silk, the synthetic threads and cloth woven out of them were eventually produced commercially and called rayon.

Rayon is actually not a synthetic fiber but a reconstituted one, yet it is one of a large family of nonwool, noncotton, and nonsilk substitutes that seems to be in everything we wear, from socks to ski jackets, from watch straps to fake furs. True synthetic materials are made from organic polymers, many of which soften if they are heated. Nylon, the first commercially successful synthetic fiber, is made into clothing, rope, and parachutes. It also is cast and molded to make zippers, gears, and bearings.

The range and variety of synthetic fibers are enormous. Dacron, a polyester, has a nonstretch quality; polyurethane elastomers stretch; Quiana, a silklike nylon, holds a crease and remains wrinkle free; carbon fibers reinforce; and high-strength aramid and polyethylene can resist bullets.

Most synthetic fibers are made into thread by melting the polymer and forcing it through tiny holes in a spinneret. A core construction technique spins a wrap of cotton or polyester around a continuous filament of polyester fibers, and then two or more single yarns are twisted together to form thread; another method textures the filament and heat-sets it. Air-entangled construction, used in heavy denim jeans, entangles fibers by passing them through a high-pressure air jet before twisting, dyeing, and winding. Monocord construction, used in the threads for making shoes, bonds nylon filaments together.

Faux furs, favored by those who object to the animal fur industry (or those who would rather not pay for the real thing), also rely on what is essentially plastic for their backing and furlike pile. In a synthetic fur, the backing is generally woven from acrylic or polyester, and the pile from long fibers of acrylic or Modacrylic, a related plastic polymer. Dyes are applied to the fibers during manufacture to produce fur lookalikes for seal, fox, leopard, and just about every member of the animal kingdom whose fur keeps them, and humans, warm.

SHEER BEAUTY

(Right) Exemplar of the many uses of nylon, women's hosiery is also a fashion statement and, encasing a pair of shapely legs, the frequent subject of photographers. Early stockings were seamed up the back by hand, but seamless hose became available in the 1940s when nylon replaced silk.

SEE ALSO

Incredible Fabrics · 84
Plastics · 90
Oil Processing · 110

SPINNING NEW WEBS

Spider silk has long intrigued engineers and scientists. Biotechnology companies are now working on a method for producing artificial fibers by inserting spider-silk genes into the cells of mammals like cows and goats. They hope to harvest silk proteins from the animals' milk. Scientists singled out the genes responsible for "dragline" silk—the tougher strands spiders use for the outer edge and spokes of their webs. While no product has yet come of this research, engineers envision using genetically engineered spider silk for sutures, parachute cords, aircraft cables, artificial tendons and ligaments, bulletproof vests, and lightweight body armor.

INCREDIBLE FABRICS

For protection in battle, Persian and Roman warriors clothed themselves in overlapping metal scales attached to linen or leather. In the Middle Ages, the Crusaders wore chain mail made of interlaced metal rings over their clothing. This flexible armor was effective, but it was still heavy, unwieldy metal.

Today's flak jackets and other types of bullet-resistant vests worn by soldiers and policemen are not the product of a blacksmith's shop. They were created by a new thread science that fabricates tough, lightweight clothing designed to protect wearers from fire, cold, and water. Fabrics can be reinforced and strengthened with carbon fibers produced from simple coal tar or by heat used to chemically change rayon or acrylic fibers. Polyester or nylon can be treated with resins that form tough, smooth protective coatings on thread. Warm, fleecelike fabric can be made from flaked and melted plastic bottles. Cloth can also be woven with glass fiber or fine metal wire.

Among the noteworthy and widely used innovations is Kevlar, a DuPont trademark for a highly crystalline polymer that is dissolved in a solvent and ultimately drawn into incredibly strong fibers used in bullet-resistant vests, military helmets, ropes, and gloves. Another well-known fabriclike material is Gore-Tex, a waterproof, windproof, and "breathable" product — a trademark of W. L. Gore and Associates, Inc. Gore-Tex fabric is actually layers of polymers fashioned into membranes and bonded to a special fabric. The membranes are full of microscopic pores, thousands of times smaller than a raindrop. They prevent water from entering but let sweat vapors leave, which allows the wearer to stay drier and cooler in warm, wet conditions.

Even more forward-looking are the so-called smart fabrics. In one design—a spin-off from materials used to keep astronauts' gloved hands warm during space walks—microencapsulated phase-changing materials (microPCMs) absorb and store heat, then release it in response to temperatures next to the skin. Another new fabric, called Holofiber and manufactured by Hologenix, LLC, is billed as a body-responsive textile. It is designed to combine ambient light energy with energy made by the body into an energy with a different wavelength altogether, which absorbs back into the body and can, for instance, increase oxygenated blood flow. In diabetics, for example, such an improvement in skin oxygenation could accelerate wound healing.

All this textile wizardry may eventually offer suits that change shape when mood or temperature changes; clothing that may be wired with locating devices; battle wear that may warn of and deactivate toxic chemicals and germs; and chameleonlike camouflage fabrics that can change colors and patterns to match the background.

VESTED INTEREST
(Right) Protected by hat and slicker, a cowboy in the rain manages chores. Ever since it became known that creatures like sea otters had waterproof fur, and that water rolls off some butterfly wings, scientists have sought ways to waterproof what we wear. What have emerged are non-breathable materials (which unfortunately retain perspiration), made of vinyl plastic compounds or other materials coated with them, and more comfortable, breathable materials that transfer moisture vapor from inside a garment to the outside.

SEE ALSO

SEWING & WEAVING

SEWING MACHINES AND KNITTING NEEDLES RELY ON THE SAME simple principle: the loop. A machine-sewn seam holds fabric together by loops of thread; knitting needles make fabric out of rows of interconnected loops. Sewing and knitting machines do the job faster, but the idea is still the same.

The sewing machine dates from 1790. In the mid-19th century Elias Howe, an American inventor, patented one that contained many features of the modern machine. Whether powered by an electric motor or a foot on a treadle, the sewing machine needs a grooved needle threaded through an eye near the point and a thread-filled bobbin rotating beneath the fabric. As drive belts push and pull the needle through the fabric, the thread in the needle loops around the thread from the bobbin and makes a tight lockstitch. Another driveshaft operates a so-called feed dog that moves the fabric along.

The knitting machine was invented in 1589 by an English cleric, William Lee. Queen Elizabeth refused him a patent because the device was a threat to hand-knitters. Now high-speed electronic knitting machines, driven by pattern software, have a computerized system that directs needle patterns and machine speed.

FROM THREAD TO CLOTH
(Below) Weaving on both traditional and automatic looms requires a warp, or lengthwise thread, and a weft, or crosswise thread. Weft yarns on a bobbin shuttle swiftly back and forth between lowered and raised warp threads, while a comblike device, or reed, spaces warp yarns evenly. A cylinder at one end of the frame keeps warp thread taut, while one at the other end holds finished cloth. Looms have 2 to 16 harnesses—framelike devices operators use to manipulate the warp and weave a variety of colors, patterns, and threads into the fabric.

SEE ALSO

Synthetic Fibers · 82
Incredible Fabrics · 84

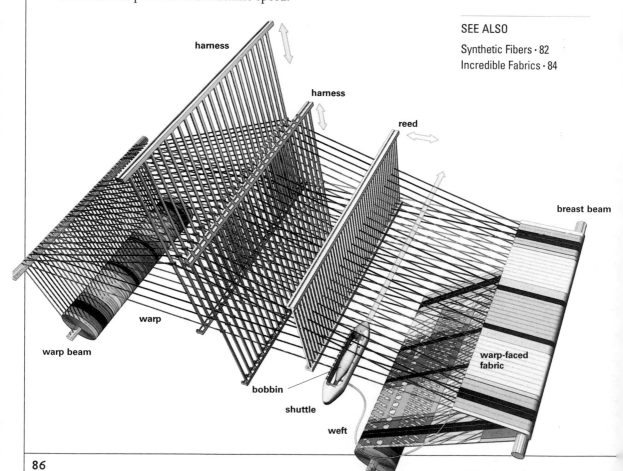

harness

harness

reed

breast beam

warp

warp beam

bobbin

shuttle

weft

warp-faced fabric

SPOOLS AND DOGS

(Right) Four spools feed a state-of-the-art sewing machine. Just like a simpler home machine, this piece of equipment has a so-called feed dog that advances the cloth as the needle, thread, and bobbin work together to make simple stitches. As each stitch is completed, the dog pushes the cloth along, rises up, then retracts as the needle comes down again. Even treadle machines, like the 1851 model named after inventor Isaac M. Singer, had feed dogs.

GLASSMAKING

W E PEER INTO GLASS TO SEE OURSELVES AND THROUGH IT to see others, for this material is both reflective and transparent, depending on how it is treated. Its basic composition can be reduced to three essential ingredients melted together at high temperatures: sand, soda, and limestone. In more elaborate chemical terms, this translates into silicates and an alkali flux, with metallic oxides added for color.

Glass is similar to most other solid materials in some ways, but on a microscopic level it lacks the orderly molecular arrangement of true solids. This disorder makes it resemble a liquid, and in fact glass has been termed "the rigid liquid," referring to its high viscosity—the property of a fluid by which it resists shape change. Glass is transparent because its atomic arrangement does not interfere with the passage of light. A glass mirror reflects because a thin layer of molten aluminum or silver is applied as backing.

The actual manufacture of glass dates back to 3000 B.C., when Egyptians glazed ceramic vessels with it. Much later, the Romans made glass for utilitarian and decorative purposes. The art of stained glass, made with metallic oxides fused into it, flourished throughout the Middle Ages.

Today glass is made in large crucibles in furnaces where the melting temperature reaches 2900°F. Skimmed of impurities and cooled, molten glass may be poured into molds and pressed, blown, cast for lenses, or floated—that is, drawn out to produce window or mirror glass. Shaped glass goes through an annealing process to fix colors and to remove internal stresses and make it less brittle. Some products that require high strength, such as glass doors or eyeglasses, are specially tempered, a rapid cooling process that is the reverse of annealing in that it induces high, permanent stress.

The cooled glass is then ground, polished, bent, laminated, or decorated. Computer-operated cutters slice the glass into different shapes, while other automated equipment grinds edges, drills holes, bevels, and inscribes. The production speed of many items is striking: Machinery can now produce over a million glass bottles a day.

BOTTLES

(Right) Glowing from their heat treatment, new glass bottles take shape on a production line. Except for automation and new materials, the manufacture of glass bottles has changed little since people in Egypt and Mesopotamia made glass vessels centuries ago.

GLASS

(Above) A glassblower in her shop in Cannon Beach, Oregon, plies a time-honored trade. The ancient Romans and Syrians, among others, learned how to make glassware by firing and melting one end of a glass tube and then blowing through it to form a bubble that could be shaped. Later, blowpipes were used to inflate a "gather" of molten glass at the end and work it into various shapes.

SEE ALSO

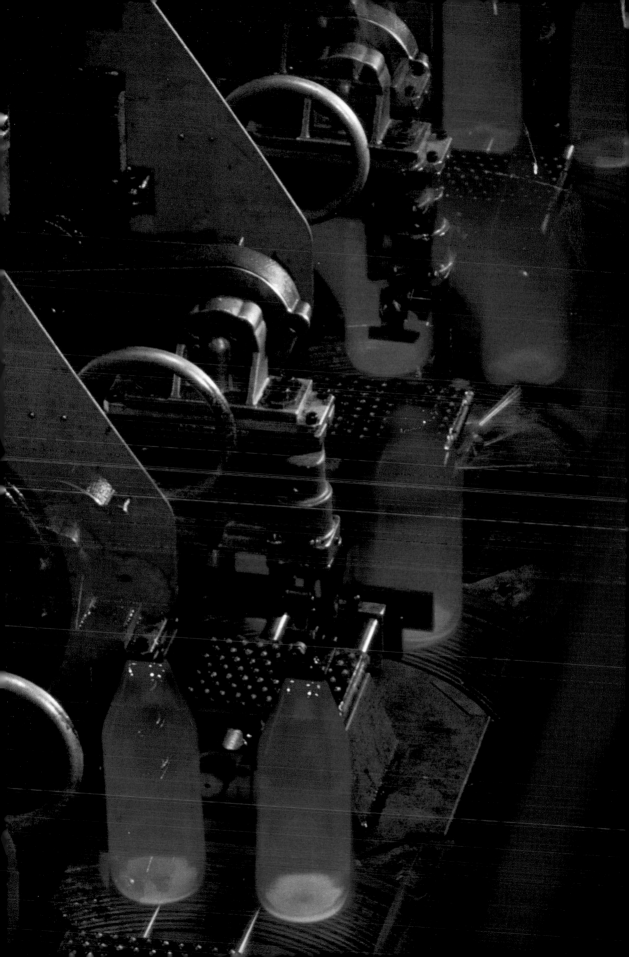

PLASTICS

I F WOOD IS ONE OF NATURE'S GREAT GENUINE ARTICLES, PLASTIC IS the welcome artificial usurper, a take-charge material with a seemingly endless capacity for changing shape and function. Most plastics are by-products of petroleum or coal, which are organic materials, but they are considered artificial because they are made rather than grown.

Plastic came into being more than a century ago when a $10,000 prize was offered to the person who could create a substitute for ivory in billiard balls. John Wesley Hyatt of New York, an inventor and entrepreneur, entered the competition. He did not win, but he did come up with semisynthetic celluloid, a mixture of nitrocellulose and camphor put under heat and pressure. Hyatt's invention was a commercial success, used in dental plates, eyeglass frames, combs, and men's collars.

Today, more than 50 varieties of plastic stand in for a host of materials, and it is hard to imagine anything spun or molded without a plastic presence. As Lucite or Plexiglas, plastic substitutes for glass in aircraft windows, car taillights, boat windshields, clock faces, and camera lenses. As Bakelite, it replaces rubber as an electrical insulator. As Corian it passes for marble, and as Styrofoam, for insulation and disposable food containers. Polyvinyl chloride (PVC) substitutes for metal in drainpipes; Teflon keeps food from sticking to skillets; nylon and other synthetic fibers replace wool and cotton.

Plastics generally start out as powders of a binder (a polymer), plasticizers, fillers (talc or glass), pigments, and other additives. One additive recently developed at Penn State, natural clay, vastly improves various properties of plastics. According to Evangelos Manias, a materials scientist at the university, clay particles are incredibly tiny, no more than three to five atoms thick; when these are dispersed in a polymer, they increase its strength and resiliency and improve its ability to keep gases and liquids in or out, a boon to food packagers.

To make plastics, the petroleum-based raw materials are compressed in molds under heat and pressure or poured into cold molds to harden. They can be squeezed between rollers or extruded through a die to be cut in lengths or coiled.

Ubiquitous and discarded by the billions, plastic bags are made from paper-thin sheets of polyethylene, then shaped, folded, and fastened into shapes meant to hold groceries, zip up food for the fridge, and haul the trash. Roughly a thousand pounds of plastic material go into making 50,000 bags. Concerns about the role of plastics in littering, however—and the fact that they can survive for generations—have resulted in their being targeted by environmentalists as a menace. Plastics do present environmental problems, but they are remarkably useful creations and they are here to stay, for better or for worse.

PLASTIC AVENUE
(Right) A couple takes a dry stroll underwater at the Detroit Zoo's Arctic Ring of Life exhibit. The submerged tunnel is made of Lucite, a tough plastic glass substitute known chemically as polymethyl methacrylate resin.

SEE ALSO

PAPERMAKING

THE OLD ADAGE "NOT WORTH THE PAPER IT'S PRINTED ON" PAYS homage of sorts to one of the most valuable, capital-intensive, and multipurpose products in the history of manufacturing.

Paper still rules the high-tech world despite repeated dreams and promises of a computerized, hence paperless society. Even "legal tender for all debts, public and private" is made from it.

The word "paper" derives from the word "papyrus," a tall water reed that the ancient Egyptians layered, pressed, and dried to make sheets for writing. Paper as we know it today was actually invented by the Chinese around the second century B.C. Made from a boiled mush of plant fibers, discarded fishing nets, and rags, primitive paper was often used as clothing and in lacquerware, kites, and currency.

Modern papermaking relies on the same general principles, except that today's sprawling mills begin with logs and employ a variety of mechanical and chemical methods to chip them, then turn those chips into pulp, an oatmeal-like slurry. Sometimes cotton rags go into the formula, and the product is a high-quality, rag-content paper. Sometimes waste paper is added in with the wood chips for the pulp, and then the product is recycled paper.

In order to make wood pulp into paper, it must be "digested"—broken down so thoroughly that individual wood fibers have separated from one another. There are a number of different pulping methods, and to a large extent these determine the character of the resulting paper. The pulp is treated by both mechanical and chemical means to remove unwanted materials that can cause discoloration. It is bleached or, more often now, treated with environmentally safe, chlorine-free agents, then beaten in machines called refiners to improve its quality and strength. The near-final paper product is rinsed and spread onto a moving belt made of wire or plastic mesh. This part of the process dries out the pulp, which is at the beginning almost 99 percent water. The moving belt shakes the water out and mats the fibers as they weave together into a thin, flat layer.

As the now drying web of paper moves down the belt, it passes through rollers and heated pressure drums that remove the rest of the moisture. Various coatings, dyes, sizing agents, and fillers are applied at this stage to impart shade and color, gloss, or other characteristics. The mass of intertwined fibers, now paper, goes through another set of metal rollers for "calendering," a process that presses and smooths it. At the end of the line, the paper is wound into rolls, cut into sheets, or trimmed by automated equipment. Packed as reels or reams, the finished product is stored on pallets in light-protected areas and eventually sorted for delivery by computerized machinery.

QUALITY CONTROL
(Right) A paper mill specialist meticulously examines what may be an understatement, given its enormous size: a sheet of paper. Before newly made paper is removed from its take-up rolls and cut and trimmed into the downsized sheets we all know, it must meet stringent requirements for color, texture, weight, and strength. Some rolls may be sold directly to a customer after inspection, while others of a different grade go through further smoothing and polishing processes and may be trimmed to various sizes.

SEE ALSO

Elements of Construction · 46
Synthetic Fibers · 82
Recycling · 94

RECYCLING

ACCORDING TO WASTE INDUSTRY ESTIMATES, AMERICANS generate more than 400 million tons of household trash a year, nearly 4.5 pounds per day per person. Included in the massive dumping are billions of batteries, razor blades, and disposable diapers, along with millions of tons of paper, glass, plastic, leaves, and grass clippings. If we were somehow able to place all our discarded beverage cans one on top of another, they would reach the moon almost a score of times.

Fortunately, recycling helps. It can be done in a number of ways, such as reusing leftovers after a manufacturing process. Steel scrap, for instance, can be used to make stainless steel, or glass scrap can be remelted to make more glass. One Japanese manufacturer, Mitsubishi Rayon, handles thousands of tons of waste plastic a year, using a heating technique to break down acrylic plastic from a complex polymer to a simpler form that becomes the raw material for more acrylic plastic. The process can convert as much as 85 percent of scrap plastic into its basic, reusable form.

Another method reclaims materials from worn-out items, using cardboard boxes and newspapers, for example, to make business cards, paper towels, and more newsprint paper. Lead batteries can be made into new batteries; steel cans make more steel; discarded glass is remelted to make new bottles. Precious metals such as gold, silver, palladium, and platinum can be recovered from printed circuit boards, plated and inlaid metals, and photographic wastes.

Old rubber can become tennis shoe soles, roofing and road construction materials, running tracks, brake linings, floor coverings, and even TDF (tire-derived fuel). Old tires are ground into bits, then chemically processed to break down their sulfur content and rearrange chemical bonds in the new rubber.

Trash has endless possibilities, as proven by a decades-long project undertaken by Indian engineer and artist Ned Chand. Chand's Rock Garden, now 25 acres in size, is a fantasy world of meandering walls, pleasant walkways, inviting arches, and intriguing sculptures. As raw materials, he used things that others had thrown away: broken china, burnt bricks, old light bulbs, unused building materials, rusting oil drums, and countless other items that used to end up in the dump.

USABLE TRASH
(Right) Sorting and sifting, a recycler gets something back from things thrown away. Recycling reduces the enormous expense of disposal and incineration; it also provides a new resource.

SEE ALSO

JUNK CARS
Our automobile-filled world creates its own special sort of junk. Stacks of crushed cars await recycling in many a municipality. A recent government study found that about 7 million cars are scrapped each year in the United States, resulting in more than 13 million tons of solid waste, largely steel and iron, but also aluminum, copper, other metals, rubber, and plastics. Presently about two-thirds of these cars are processed to recover scrap materials. Steel scrap can be melted down and used again in a number of applications. Even the fluids in an abandoned car—antifreeze, fuel, and oil—if they are promptly and carefully drained, can be recycled for future use.

CHAPTER 6

MANUFACTURING

I F YOU HAVE GREAT TALENTS, INDUSTRY WILL improve them," English portrait painter Sir Joshua Reynolds once said. "If you have but moderate abilities, industry will supply their deficiency." His remarks came during a 1769 address to students of the Royal Academy, and his use of "industry" referred to diligence in any pursuit. But it also could have applied to the industrial revolution, which introduced power-driven machinery to late 18th-century England and forever changed the workings of the factory system. Today, the industrial world employs the modern tools of technology and science to manufacture products, extract value from natural resources, and create new versions of the tools themselves. Increasingly, manufacturers are relying on information technology to help them improve efficiency and become more competitive in the global market.

Coal-mining equipment glows at dusk.

AGRICULTURE

Without doubt, Cyrus McCormick's mechanical reaper of 1831—a horse-drawn, wheeled rig that gathered and cut bunches of grain stalks—was the most important mechanical advance in farming at the time. What has followed is equally impressive, a technology-rich revamping of the way agriculture is done.

Seed-handling machines transfer seed from hoppers to drill boxes and planters on streams of air. Combine harvesters that cut fruiting heads, thresh, and clean grain now come equipped with electronic fuel systems, ergonomic cab designs, and sensors that read changing terrain and automatically level the machine. Corn pickers pick, husk, and send corn to shellers that remove the kernels; cotton pickers twist fiber right off the bolls. Cows are milked robotically.

Even baling hay has improved. Tractor-pulled balers now have self-feeding pickups, automatic knotters, and bale throwers that fling the bales into wagons. One such tosser, the Bale Flipper, works without a tractor to load bales of hay onto a cradle trailer. It automatically loads a bale into one cradle, then moves back to fill the next. When the last bale is loaded, the apparatus rolls off the back of the trailer and an electronic switch automatically turns off the motor.

A COMBINE

(Below) This versatile rig does just what its name implies. Moving through a field, it combines the tasks of the harvest, simultaneously reaping, threshing, and cleaning grain. The tined reel sets up the grain heads, and a cutter bar slices them. An elevator conveys them to the threshing cylinder, which separates the grain and drops it onto a vibrating pan. The grain then goes to a sieve, and an auger and elevator carry it to the holding tank. The rear beater sends the threshed stalks to straw walkers that eject them for baling.

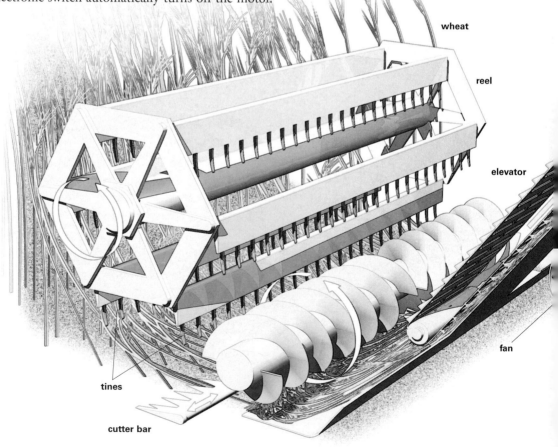

wheat

reel

elevator

fan

tines

cutter bar

HARVESTING

(Right) A field-mowing combine harvester spews out the cereal grain for a nation's breakfast table.

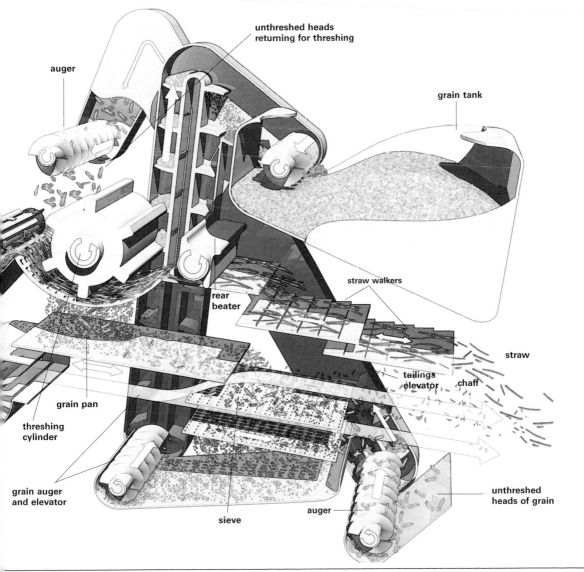

auger

unthreshed heads
returning for threshing

grain tank

straw walkers

rear
beater

straw

tailings
elevator

chaff

grain pan

threshing
cylinder

grain auger
and elevator

unthreshed
heads of grain

sieve

auger

BIOTECHNOLOGY

IT TASTES LIKE CAULIFLOWER WHEN EATEN RAW, BUT IT IS MILDER, sweeter, and more like broccoli when cooked. Say hello to the broccoflower, a chartreuse cross between the two members of the family Cruciferae. First marketed ten years ago, the broccoflower—along with FLAVR SAVR tomatoes and SuperSweet onions—is a product of food biotechnology, a broad-ranging science employing a host of techniques to improve the things that we eat. Scientists can genetically alter crops by introducing new genes, thereby shielding them from disease, spoilage, and insects or making them grow faster, with less dependence on chemicals. Crops also can be gene-goaded into faster ripening rates and higher yields, or they can be endowed with better nutrients and more flavor.

One important example of biotechnology is selective breeding, an agricultural standby that involves selecting plants and animals with desirable traits and breeding them under controlled conditions. Selective breeding concentrates on entire organisms with complete sets of genes, while genetic engineering focuses on a few gene transfers. The two techniques have given us seedless bananas, beardless mussels, leaner animals, and virus-resistant squash. Synthetic Bovine Growth Hormone (BGH), a genetically engineered drug given to cows, increases milk output by augmenting a cow's own natural production of the hormone. More than half the cheese produced in the United States is made with chymosin, a biotech preparation that does away with the need for rennin extracted from calves' stomachs. Tomatoes now made tough enough to ripen on the vine and then ship without spoiling are a plus when you consider that commercial tomatoes are generally shipped green and ripened with ethylene gas.

But there are, as in all cases of genetic manipulation, red flags. While proponents argue that the techniques are safe—indeed, the FDA recently concluded tentatively that milk and meat from cloned animals are safe to consume—opponents argue that gene splicing might instill lethal allergens in foods or deprive food of its nutritional value. As one analysis of the situation put it, "Whatever their interpretations of the costs and benefits, both proponents and opponents agree that biotechnology has the potential to fundamentally change how food is produced in the future."

BOTANICAL DUPLICATE

(Right) Cloned plants repose in their test tubes, the product of manipulated genes. Chinese scientists have cloned large quantities of the rare Venus flytrap, which eats insects. Others have been trying to put the scent back into flower varieties that have lost it in the scientific quest for bigger blooms.

SEE ALSO

FARMING BY SATELLITE

Another on-the-farm technology today is precision agriculture, by which satellite imaging of field conditions helps to improve crop yields. Weeds can be mapped with global positioning system (GPS) techniques. The light beams of sensors analyze the ability of seed furrows to accommodate the flow of air and water. Tractor-mounted computers and satellite connections identify variations in nitrogen and pH levels, anticipating fertilizer and pesticide needs within feet of a tractor's position. Right now dozens of satellites are orbiting the Earth in several different paths, sending signals to help map out the mosaic of crop-producing fields in this and other countries.

AQUACULTURE

AQUACULTURE, THE CULTURING OF FISH, SHELLFISH, AND PLANTS in a controlled area of water, is a rapidly growing agribusiness, larger even than veal, lamb, and mutton combined. According to the U.S. Department of Agriculture, 20 percent of the fish consumed in the United States is now raised on farms, with catfish, tilapia, salmon, trout, crawfish, and shrimp among the leaders.

Fish farms may be earthen ponds, tanks, rafts, troughlike raceways, net pens, suspended cages, or bottom nets. The "seeds" are fingerlings, very young mollusks, or eggs, which are raised on commercial feed or natural organisms grown through water-fertilizing techniques.

Potentially polluting and disease-bearing waste accumulates in any fish farm, and aquaculture farmers dispose of it through elaborate water-circulation systems. Some filtration systems harness bacteria that convert ammonia, secreted by fish through their gills, into nitrates that can be flushed from tanks. As an additional precaution, stocked fish may be treated with antibiotics and other chemicals, a practice that invariably raises concerns.

One state-of-the-art technique, gene splicing, has been found to produce catfish, salmon, and trout with more resistance to disease than their wild counterparts. The genetically modified fish grow faster and develop a stronger immunity to freezing in winter. Some observers worry, though, that as desirable as these characteristics are in a fish population, the genetically modified (GM) breeds pose great risks to wild populations. When the GM fish escape, which farmed fish inevitably do, they will breed with wild fish and produce modified offspring, potentially compromising the natural gene pool.

Some fish, such as salmon, which hatch in fresh water and then swim to the ocean in the wild—must be transferred from fresh to salt water. Farmed salmon can suffer a trauma called osmotic shock. Handling them with special water and feeding them a formula can solve the problem.

Shellfish farming is also fraught with environmental hazards. Growing oysters, for example, involves fertilizing female eggs, which are up to 90 microns in diameter, with male sperm, about 10 microns long. Large quantities of phytoplankton are fed to larvae and older oysters, which must be transferred from one bed to another in order to safely accommodate the oysters' various stages of growth.

FISH HIGHWAY

(Below) A little fish flits through a hose at an Idaho fish farm, while its relatives in the wild may ride a tide onto shore or travel upstream against a strong current. Unlike conventional farms, which generally consist of vast tilled fields arranged in familiar patchwork-quilt patterns, water-filled fish farms prove the ultimate in container-growing.

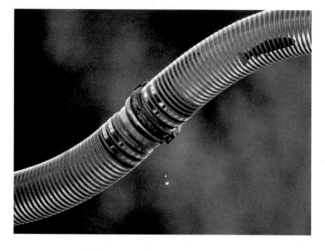

SEE ALSO

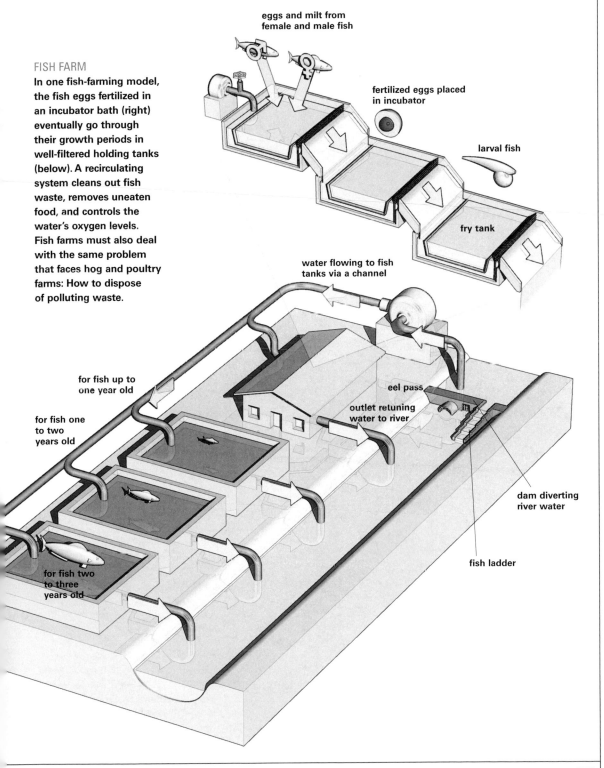

FISH FARM

In one fish-farming model, the fish eggs fertilized in an incubator bath (right) eventually go through their growth periods in well-filtered holding tanks (below). A recirculating system cleans out fish waste, removes uneaten food, and controls the water's oxygen levels. Fish farms must also deal with the same problem that faces hog and poultry farms: How to dispose of polluting waste.

eggs and milt from female and male fish

fertilized eggs placed in incubator

larval fish

fry tank

water flowing to fish tanks via a channel

for fish up to one year old

for fish one to two years old

for fish two to three years old

eel pass

outlet retuning water to river

dam diverting river water

fish ladder

HYDROPONICS

CONSIDERING THE VULNERABILITY OF SOIL TO DISEASES, PESTS, changing weather, and inadequate nutrient supply, it is no wonder that farmers long for better control of their plants and sometimes even wish they could control growing conditions. One remedy is the science of hydroponics: Plants are grown without soil in nutrient-enriched solutions, with their roots anchored in porous non-soil materials. Widely used in botanical research, hydroponics also grows vegetables, fruits, flowers, and herbs.

The concept is not new. Water gardens thrived along Africa's Nile River thousands of years ago. During the Second World War, the U.S. Army grew vegetables hydroponically on infertile Pacific islands. Crop yields can exceed the success rate of dirt farming, but large-scale soilless farming remains confined to out-of-season greenhouse plants and to areas that have limited arable land.

Soilless culture begins with water enriched by the same balance of nutrient salts found in soil. These nutrients are absorbed by plant roots, supported by all manner of materials that retain air and water, including sand, gravel, glass wool, rock wool fiber, and stone.

A variation of this process is aeroponics, by which plant roots are suspended in a chamber or a bag. Humid air provides the proper environment, while a spray mist of nutrient solution keeps the roots moist and nourished. Almost no water is lost through evaporation, and roots absorb much oxygen, increasing metabolism and the rate of growth as much as ten times over that in plants growing in soil.

NASA's interest in growing food in space has led to the development of a medium made from zeolite, a common mineral that works as a "molecular sieve," storing and time-releasing nutrients. Additive-enhanced zeolite creates conditions almost comparable to those of the soilless "soil" in conventional hydroponics.

A HYDROPONIC GARDEN

(Right) An artificial environment provides all of the ingredients essential for plant growth: oxygen, light, heat, water, nutrients, and carbon dioxide. Inside the protective confines of a greenhouse, plant roots absorb nutrient salts from enriched water. An inert and soil-free porous medium and plastic mesh anchor the roots in the water. Sun or artificial lighting provide adequate light. By delivering nutrients directly to the roots, a hydroponic system ensures that energy ordinarily used to produce long roots goes directly into growing a larger plant.

SEE ALSO

SLUDGE-BUSTERS

(Left) Contained nutrient solutions used in hydroponics do not pose a threat, but runoff from fertilized soil pollutes aquifers and presents a clean-up challenge. At an experimental greenhouse in Providence, Rhode Island, sewage is piped into vats, along with plants, bacteria, snails, and fish with an appetite for filth. In a few days, the cloudy water will be as clear as the sample in the other flask, virtually odorless, and ready for discharge. Bacteria have also been harnessed by scientists to devour oil spills. They break down the oil in much the same way that ordinary bacteria cause dead animals and plants to decompose and change into soil-enriching substances.

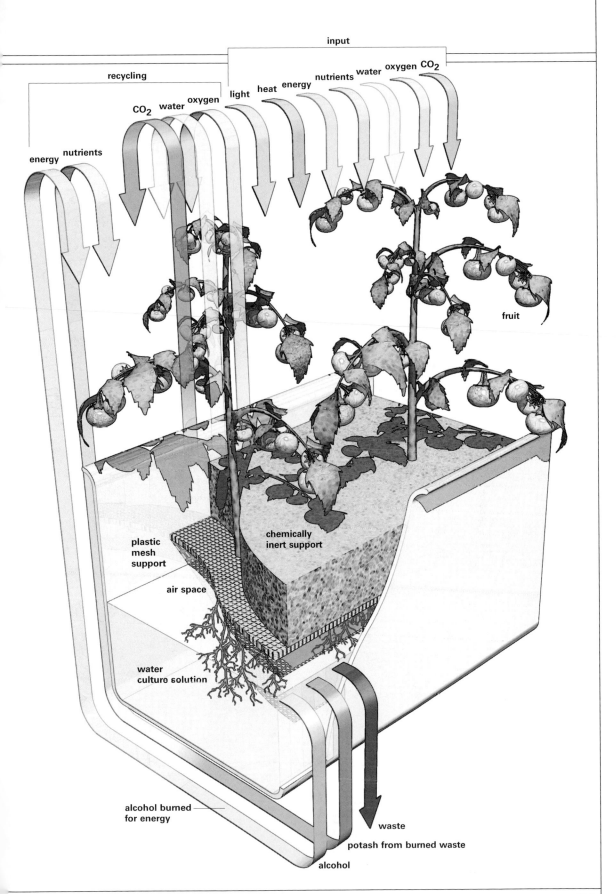

input

recycling

CO₂   water   oxygen

light   heat   energy   nutrients   water   oxygen   CO₂

energy   nutrients

fruit

plastic mesh support

chemically inert support

air space

water culture solution

alcohol burned for energy

waste

potash from burned waste

alcohol

STEELMAKING

EW INDUSTRIES CAN MATCH THE MASSIVE, COMPLEX EQUIPMENT and the awesome spectacle of steelmaking. From raw and dirty lumps of iron ore to sheets of shiny, corrosion-resistant stainless steel, the process is a meld of intense heat, blasts of high-pressure air, molten metal, violent boiling and bubbling, arcs of electricity, and the din of forges and rollers. First mass-produced in the 19th century, steel is a mainstay of ships, automobiles, and skyscraper frames.

Steel is an alloy of iron and a tiny amount of carbon and other elements, a mix that causes the iron atoms to bind together tightly and produce a material that is even harder and tougher than iron. Three key pieces of equipment in steelmaking are the blast furnace, the Bessemer converter, and the open-hearth furnace.

To create steel or to make cast or wrought iron, the iron is first extracted by smelting oxide ores mixed with coke (a form of carbon) and limestone in a blast furnace—a towering, cylindrical stack where a blast of air fuels combustion. In molten form and full of impurities (pig iron), it can be made into cast or wrought iron.

Pig iron turns to steel when it is refined in a Bessemer converter, which uses hot compressed air or an open-hearth furnace to remove impurities. Stainless steel, a long-lasting alloy containing at least 10 percent chromium, is generally produced in an electric-arc furnace, which produces batches of molten steel known as "heats." Carbon electrodes supply energy to the furnace's interior, where iron is melted so that it can be mixed with stainless steel scrap, chromium, and other elements, such as nickel and molybdenum. The mix is cast into ingots or a slab and then hot-rolled, cold-rolled, or forged into final form.

SPARKS FLYING
(Right) Sparks and smoke define the process of turning molten metal into steel. Melted iron seethes at a temperature of 2800°F, when it will be poured from a cauldron into a steelmaking furnace. Manufacturers can use Bessemer converters, or electric-arc or open-hearth furnaces.

MAKING STEEL
(Below) When workers dump iron ore, limestone, and coke into a blast furnace, the coke and blasts of air fuel combustion. From the furnace, molten pig iron goes to a converter charged with oxygen to burn out impurities and convert the iron to steel. The next step involves removing the molten steel and tipping the converter to drain off slag.

SEE ALSO

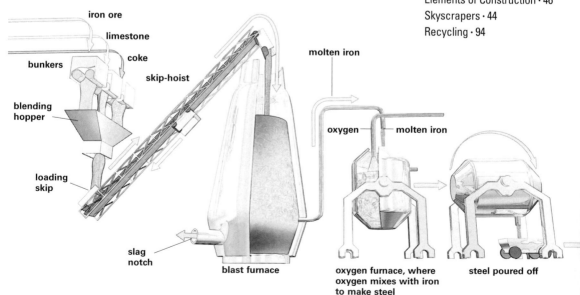

iron ore

limestone

coke

bunkers

skip-hoist

molten iron

blending hopper

oxygen   molten iron

loading skip

slag notch

blast furnace

oxygen furnace, where oxygen mixes with iron to make steel

steel poured off

COAL MINING

THE WORLD'S VAST STORES OF COAL WERE PROBABLY LAID DOWN some 300 million years ago during the Carboniferous period, when the remains of plants decomposed and compressed to form the hard, black substance we burn as fuel. The U.S. alone has trillions of tons of this organic rock, and it is enough to take us, at the current rate of consumption, well into the 24th century.

One way coal is extracted is by excavating through strip mining, a hotly debated process that removes surface material from mountaintops and other areas to expose coal seams or beds. The equipment is mammoth: A typical truck's bumper might be as high off the ground as a worker's helmet, and the load carried can be about 250 tons a trip. Critics of this type of mining argue that it not only mars the landscape but produces acidic runoff and loads streams and other nearby water with debris.

Underground mining is the other major method of extraction. Entering by tunnels or shafts dug into the ground, miners had to muscle out their coal carts. They were paid by the tons of coal they loaded. Today, new roof-bolting technologies are used, and mines are drained, ventilated, and fitted with mechanical and electronic equipment to make them safer.

A typical underground mine is excavated by using what is called the room-and-pillar method. Rooms generally 20 to 30 feet wide are cut into the coal bed, leaving huge pillars, or columns, to support the roof and control the airflow. The cutting is done by a machine called a continuous miner, which obviates the need for blasting and drilling.

When the cutting reaches the end of the line, workers begin "retreat mining," a process by which they pull as much coal as possible from the remaining pillars—until the roof collapses.

Most coal—about 65 percent of that used in the United States—travels to where it is needed by railroad, truck, barge, and conveyor, including pipelines through which coal slurry travels directly from mines to power plants. Coal is often pulverized for power generation, but coal gasification can also produce a gas used to fire a turbine to produce electricity. Coal liquefaction converts the coal into liquid fuel, used in the manufacture of plastics and solvents.

OPEN PIT MINE

(Below) Resembling an ancient amphitheater, an open pit, or surface, mine like this one works best when the coal or metal ore is relatively close to the surface. Open mines provide more than 60 percent of U.S. coal.

SEE ALSO

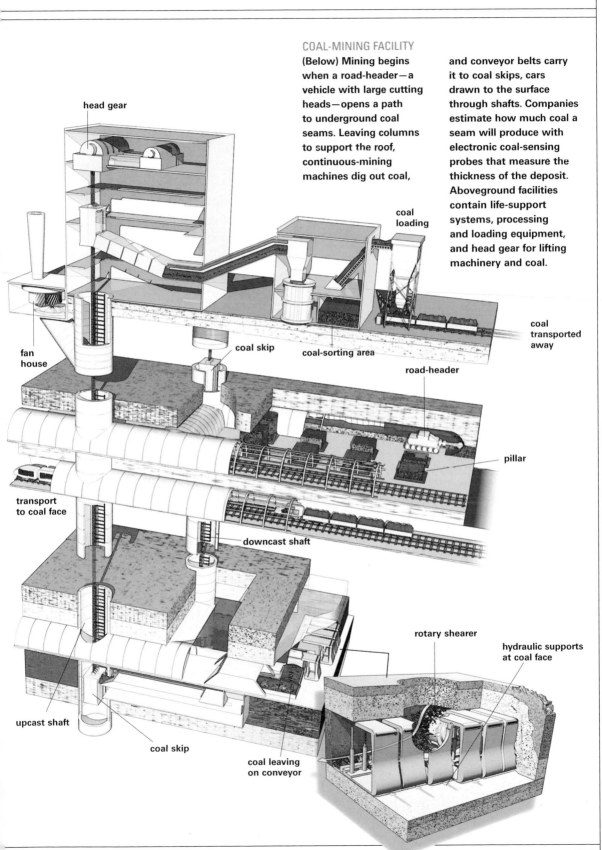

COAL-MINING FACILITY

(Below) Mining begins when a road-header—a vehicle with large cutting heads—opens a path to underground coal seams. Leaving columns to support the roof, continuous-mining machines dig out coal, and conveyor belts carry it to coal skips, cars drawn to the surface through shafts. Companies estimate how much coal a seam will produce with electronic coal-sensing probes that measure the thickness of the deposit. Aboveground facilities contain life-support systems, processing and loading equipment, and head gear for lifting machinery and coal.

head gear

coal loading

coal transported away

fan house

coal skip

coal-sorting area

road-header

pillar

transport to coal face

downcast shaft

rotary shearer

hydraulic supports at coal face

upcast shaft

coal skip

coal leaving on conveyor

OIL PROCESSING

A COMPLEX MIXTURE OF HYDROCARBONS THAT FORMED FROM THE organic debris of long-dead plants and animals, petroleum is the oily, flammable liquid of power and petrochemicals. Pooled deep in the ground, within layers of rock under intense pressure, it is extracted by a drilling rig, a series of rotating pipes supported by a derrick.

But first, oil must be found. Sensing devices guide drills to where oil is located, and the underground reservoir is tapped. Oil tends to burst out explosively, but today's technologies have made such spectacles a thing of the past. Drilling rigs probe to depths as great as 25,000 feet and pump a flushing mud into the ground, carrying debris to the surface and preventing such eruptions.

After a well is drilled, the oil usually flows up under its own pressure and into a separator, where gravity helps remove it from the briny water, natural gas, and sand that comes up along with it. The cleansed crude oil then travels through an array of pumps, compressors, and dehydration towers to make it suitable for refining; at the refinery, petroleum undergoes fractional distillation, which produces gasoline, kerosene, diesel and fuel oils, lubricants, and asphalt.

In the early days of prospecting for oil, a reservoir often was discovered by chance, but now geologists examine soil and rock with the help of satellite images, gravity meters, and magnetometers, which analyze changes in the Earth's gravitational and magnetic fields that suggest the presence of oil.

Electronic equipment nicknamed "sniffers" pick up the smell of hydrocarbons, and seismometers assess shock waves deliberately created, for they help detect hidden oil deposits. Ultrasound waves provide color-coded seismic data with images of rock formations in the ground or seabed.

BRIGHT LIGHTS
(Below) A gleaming petrochemical plant and its dark towers dominate a landscape. The stuff of an endless range of products from plastics to pesticides, detergents, nitrogen fertilizers, and explosives, petrochemicals make up a large group of organic and inorganic compounds spawned by oil and natural gas.

SEE ALSO

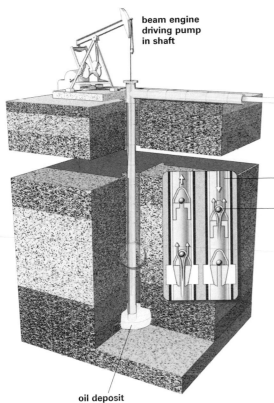

beam engine
driving pump
in shaft

oil in
pipeline

pump

ball valve

oil deposit

FRACTIONAL DISTILLATION

(Below) The fractional method refines crude oil by using the distinctive boiling and condensation points of different hydrocarbons to separate them out. Crude oil vapor enters at the bottom of a fractionating tower. The heated vapor rises, cools, and condenses on trays at different levels—the lightest fractions with the smallest molecules at the top. Catalytic cracking in a reactor breaks large molecules into more valuable small ones. A vacuum still lowers the boiling point of the heaviest, unvaporized oil.

PUMPING OIL

(Left) Activating a system of ball valves near the bottom of an oil shaft, a beam engine dips and rises to lift oil from porous rock 500 to 25,000 feet deep in the ground. On the surface, pipelines carry the oil to processors.

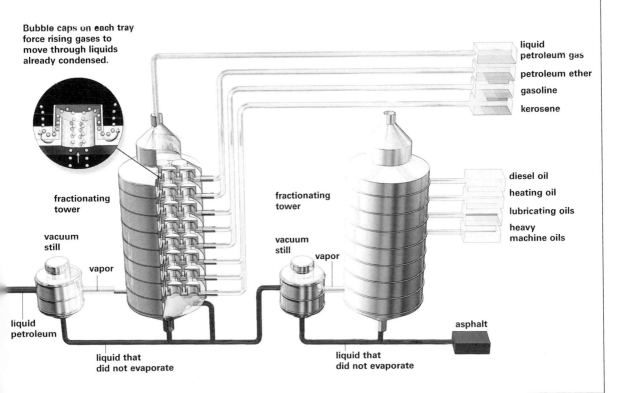

Bubble caps on each tray force rising gases to move through liquids already condensed.

liquid petroleum gas

petroleum ether

gasoline

kerosene

diesel oil

heating oil

lubricating oils

heavy machine oils

fractionating tower

fractionating tower

vacuum still

vacuum still

vapor

vapor

liquid petroleum

asphalt

liquid that did not evaporate

liquid that did not evaporate

INDUSTRIAL ROBOTICS

NDUSTRIAL ROBOTS ARE TRUE TO THE ORIGIN OF THEIR NAME: "ROBOT" derives from the Czech word *robota,* which means work, or slavery. Task-oriented machines guided by mechanical and electronic means, robots perform functions ordinarily done by humans and are especially useful when the job is boringly repetitive, dangerous, or heavy-duty, like polishing door handles or helping police dispose of bombs.

Contrary to sci-fi films, these factory work-hands do not physically resemble humans, though they may have mechanical "hands" and "wrists" with segmented gripping "fingers." They run on high-efficiency electric motors, hydraulic systems, compressed air, sensors, and solenoids, current-carrying devices that convert electrical into mechanical energy. Robots do not think for themselves; directions for what they can do and how to do it come from embedded microprocessors and auxiliary computers. Robots come in a variety of sizes and nondescript shapes, and many configurations allow for different axes of movement, or "degrees of freedom," as robot experts put it. Robots flex their steel "arms" while ranged along an assembly line, or they do incredibly detailed work while mounted unobtrusively on a table or wall, or hanging from the ceiling on a gantry. Some are mobile, like those used in underwater and space exploration, and may be controlled remotely by a human operator.

All this "technology of mobility" begins with actuators: the motors, pneumatic systems, solenoids, and so on that enable the robot to pivot, reach, gyrate, grip, and grab in seemingly limitless ways. The manipulative arm of a robot is capable of moving up and down, in and out, and side to side. It can also perform various "wrist" movements, which include rotating clockwise and counterclockwise.

Robots can be set up to do their designated tasks or series of tasks by a "teach-and-repeat" method. An operator or programmer employs a portable control device to teach the robot its job manually—for example, walking the robot's arm through the various positions. The position and functional data obtained are then programmed into the robot's computer system for retrieval when needed. Robots also learn by being fed movement data and other essential information, prewritten as a computer program. In that case, the information to be retained is transferred to the robot either directly from a control room or through storage devices.

To get the robot to go about its business, the programmed instructions are retrieved, and the computer or embedded microprocessors switch on the motors and other actuators that translate the coded data into robot motion. If it becomes necessary to alter the robot's mission, it is simply reprogrammed, using either online or offline techniques that transfer new data to the computer via storage disks or telephone modems.

ROBOT AT WORK

(Right) A robotic arm, programmed to test a piece of equipment, can spot, test, and discard defective parts precisely. While such a set of tasks may once have been performed by a human worker, over and over, a robotic worker can perform the same mechanical sequence without flagging, each iteration the same as the last.

SEE ALSO

CHAPTER 7

ENTERTAINMENT

Howard WE DEFINE IT AND WHEREVER WE seek it out, entertainment serves as humankind's leavening agent and safety valve. Over the years, many instruments have remained the same in appearance and function—Chopin would have little difficulty at the keyboard of a modern piano—but others have been transformed and improved, and many new technologies have been created. Recording media are now essential elements in the entertainment world, and the way they process information changes swiftly. Cameras and sound recordings are now digitized. Radios come microsize. Music can be made electronically. Video games let us sink and raise the *Titanic*. From the new satellite-based radio networks to the ever more exciting roller coaster, entertainment and its accoutrements have come under the powerful influence of modern technology.

Rock concerts today are feasts of electrified sound, sight, and sensation, amplified to reach thousands and recorded for millions more.

MAKING MUSIC

USIC, AS THOMAS CARLYLE SAID, MAY WELL BE THE "SPEECH of angels," but it is also the product of human hands, throat, mouth, lips, and feet. It is the result of someone manipulating the vibrations and the increases and decreases in air pressure that are at the root of all sound. Everything that makes sound sets up vibrations that disturb the air and, in turn, change the air pressure against our eardrums. The eardrums vibrate, and a sound message is carried to the brain, which differentiates between cacophony and an angelic choir. Playing a musical instrument transmits the vibrations that move through the air as audible waves.

Sound's loudness or softness depends on the amplitude of sound waves, while its pitch (highs and lows) depends on their speed of vibration. Humans cannot hear sounds having less than 30 or more than 40,000 vibrations per second. Loud, sharp sounds, such as rifle shots or police whistle shrieks, are at high frequency; they create considerable compression in the air, while a brass horn emits waves of lower frequency. Music is composed of notes, which differ from one another in their pitch, and what we know as a musical tone is actually a sound of definite maintained frequency.

The instruments that play the notes create vibrations differently. Stringed instruments create sounds when tightened strings are vibrated. Blowing into a woodwind vibrates a thin, flat reed that makes the air inside the instrument vibrate, producing notes that can be changed by opening and closing holes. Percussion instruments send out vibrations when a tightly stretched skin is struck with sticks, hands, or a mallet.

THE GRAND PIANO
(Right) Derived from dulcimers, and preceded by clavichords and harpsichords, a concert grand piano uses the same hammer action invented by Cristofori (1655–1731) for an Italian harpsichord.

SOUNDING THE TRUMPET
(Below) Pressing or releasing a piston on a trumpet causes a valve to open or close an extra section of tubing. In the "up" position, the valve shuts off the loop attached to it, and the air goes straight through. In the "down" position, the valve opens the loop and diverts the air through the extra section. By blowing with tensed lips a musician can create different notes.

SEE ALSO

Radio · 120

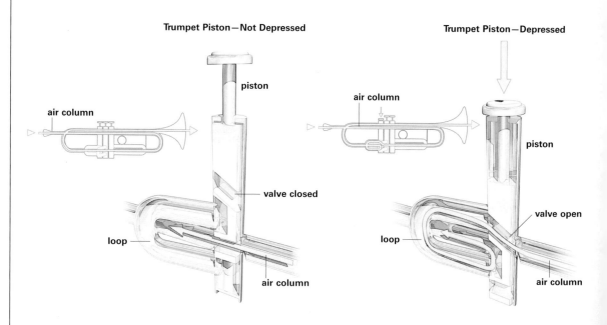

Trumpet Piston—Not Depressed

Trumpet Piston—Depressed

piston

air column

air column

piston

air column

valve closed

valve open

loop

loop

air column

air column

SENDING SIGNALS

THE MUSIC AND THE VOICES WE HEAR COMING FROM OUR conventional radios have traveled great distances and in many guises. They begin at a broadcasting station, where sound waves vibrate the diaphragm of a microphone, which turns the acoustical energy into a weak electrical signal. That weak signal is amplified and then added to a carrier wave so it can be broadcast. Each radio station is assigned a carrier wave with a different frequency.

An antenna on top of the radio station building beams out the audio-carrying radio waves—faster and more powerful than audio signals alone—at the station's assigned frequency. The distance the radio waves travel is determined by their frequency and by electrical conditions in the atmosphere through which they travel.

Waves that have left the station's transmitter are picked up by the antenna in your radio. A tuner then matches the receiver to the station's transmitting frequency, which is the way you select the program or channel you will listen to. The radio wave signals weaken as they travel across such a distance. Your radio instrument turns them into electrical signals, which are amplified and then turned into audio signals. These are sent to the radio's speaker, where the electrical waves are converted back into sound and amplified again.

Waves at different frequencies require different sorts of relay systems in order to travel far from their source. Indirect, or short, waves reflect between the sky and Earth's surface. Capable of traveling the farthest around the world, they are essential to international communication.

BOUNCING WAVES

(Right, below) The path and distance radio waves travel depend on their frequency and on electrical atmospheric conditions. Indirect—or short—waves reflect between the sky and Earth's surface. FM radio and TV signals, surface or direct waves, nearly parallel the Earth and require relay towers. VHF waves travel by line of sight, and medium waves bounce off the ionosphere.

SKY EARS

(Below) An array of radio telescope receivers in New Mexico gathers and amplifies distant signals for astronomers.

SEE ALSO

SATELLITE RADIO

(Left) Unknown until a decade ago, satellite radio now draws millions of listeners who pay a fee for coast-to-coast, static-free news and entertainment. Programming is digitized and transmitted from ground stations to proprietary satellites positioned to provide continent-wide reception. Signals bounce from satellites back to portable radio receivers installed in homes or cars. The receivers decode the signals for the listener, sending the audio along with song titles, names of artists, and other information, all displayed in text on the receiver's screen. Aside from multichannels generally free of commercials, satellite radio networks allow a driver to cross the country without worrying that the signal from her or his favorite radio station will fade.

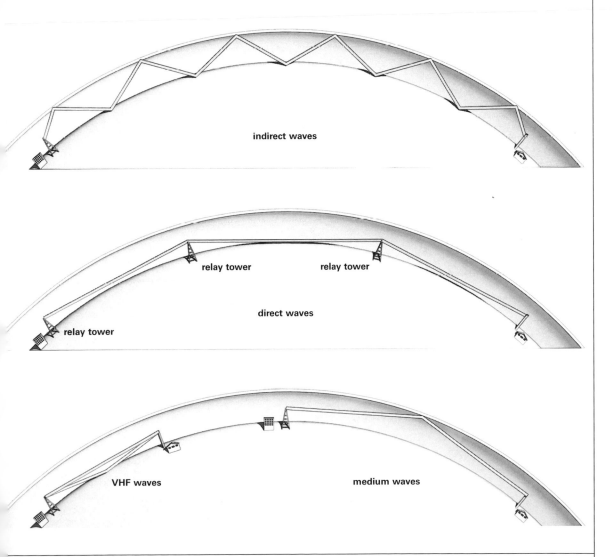

indirect waves

relay tower   relay tower

direct waves

relay tower

VHF waves   medium waves

RADIO

UGLIELMO MARCONI MAY HAVE INVENTED THE WIRELESS RADIO, but his feat drew on the research of Heinrich Rudolph Hertz, the German physicist who first demonstrated the existence of radio waves. Hertz proved that the waves could be reflected and refracted, as light is, and that they could be sent through space. He gave us hertz and megahertz, units that are used to measure the frequency of electromagnetic radiation, which includes radio waves. The units (MHz) correspond with a station's number on the radio dial.

Radio waves are made to carry signals by changing, or modulating, the waves. Some radio stations send their signals by changing the size, or amplitude, of the radio waves. These AM (amplitude modulation) stations broadcast on frequencies that are measured in thousands of cycles per second, or kilohertz. Other stations broadcast by making small changes in the frequencies of their radio signals. These FM (frequency modulation) stations are assigned frequencies in millions of cycles per second, or the megahertz range. Television makes use of both kinds of waves, with pictures carried by an AM signal and sound by an FM signal.

When a station is selected on a radio, a tuning circuit picks one and tunes out all the others by permitting current to oscillate at a single frequency. The two conducting plates of a capacitor, or condenser, store energy as electricity, while a coil to which they are linked stores energy as a magnetic field. The magnetic field collapses and sends an electric current to recharge the capacitor, which discharges again through the coil, instigating an oscillating current of one frequency. Essentially, the capacitor blocks the flow of direct current while allowing alternating and pulsating currents to pass.

MUSIC MASTER

(Right) Equipped with headset and microphones, a disc jockey spins music into the radio waves that will carry it to distant locales via a central transmitter.

MAKING WAVES

(Below) Radio waves demonstrate characteristics associated with their frequencies. Longer wavelengths have lower frequencies, for instance. Impressing sound waves onto radio waves involves amplitude modulation (AM) or frequency modulation (FM). In AM, the signal is carried by varying the amplitude of the radio wave; in FM, the signal is carried by varying the frequency along with the sound signal.

SEE ALSO

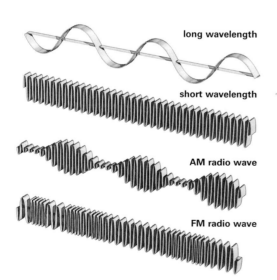

long wavelength

short wavelength

AM radio wave

FM radio wave

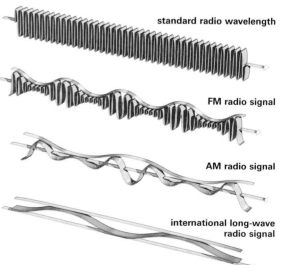

standard radio wavelength

FM radio signal

AM radio signal

international long-wave radio signal

DIGITIZED MUSIC

SOME BIRDS THAT HAVE THE RIGHT VOCAL CORDS AND THROAT formation can imitate spoken words with a fair degree of accuracy, but it takes electricity and electronics to reproduce a broader array of sounds. In the now almost extinct phonograph, a diamond stylus, or needle, played back sound vibrations that had been scribed through a microphone into grooves on a vinyl record. As the record turned, the needle picked up mechanical vibrations, converted them into electrical signals corresponding to peaks and troughs in the sound waves, and sent them to loudspeakers that converted the signals to exact replicas of the original sound. Sound's electrical counterparts may also be embedded magnetically on a specially coded plastic tape; the tape is run past a recorder's electromagnetic playback head, which sorts out the coded sound and allows it to be amplified and sent to the speakers.

What has revolutionized music storage and music making is the conversion of sound from its vibratory, analog-wave state into a digital format of numbers, and then back again to an analog wave that can be amplified and heard through speakers. The compact disc (CD), along with its computer counterpart, the CD-ROM, are read by a thin laser beam that probes sequences of microscopic pits and flat regions for binary codes that will be reconverted into sound.

If the CD was a breakthrough in sound recording, the blending of computers, the Internet, and MP3 players has totally transformed the way people acquire, listen to, and share music. The MP3 is an audio compression format that shrinks digital files without affecting the resulting sound quality.

Music can also be produced entirely by electronics. A synthesizer, for example, is a keyboard system of waveform generators and computer connections. It simulates the sounds and overtones of a variety of instruments by creating the appropriate electrical sound signals.

Another system, known as MIDI, for Musical Instrument Digital Interface, has virtually erased the boundary between the real and the artificial. MIDI allows synthesizers to link up to other synthesizers, computers, and to sounds from different instruments. The electronic keyboard—along with hardware, software, controllers, sequencers, power amps, and speakers—lets a musician design and layer sounds, dub and overdub, fill, edit, and play. Portable MIDI keyboards allow a musician to use them like high-tech scratch pads, improvising sound.

Under MIDI's direction, an entire music studio's output can be controlled, and a single musician—a keyboardist, for instance—can become a one-person band. The future may even hold "hyper-instruments," electronic components that understand the performer's intentions and enhance musical expression.

A CD REVEALED

(Below) Scanned by an electron microscope, the cracked surface of a compact disc displays the musical layer below, at center. Made of plastic, the disc has been pressed with a series of fine depressions, or notches, which represent a digitized musical signal capable of being read by a laser.

SEE ALSO

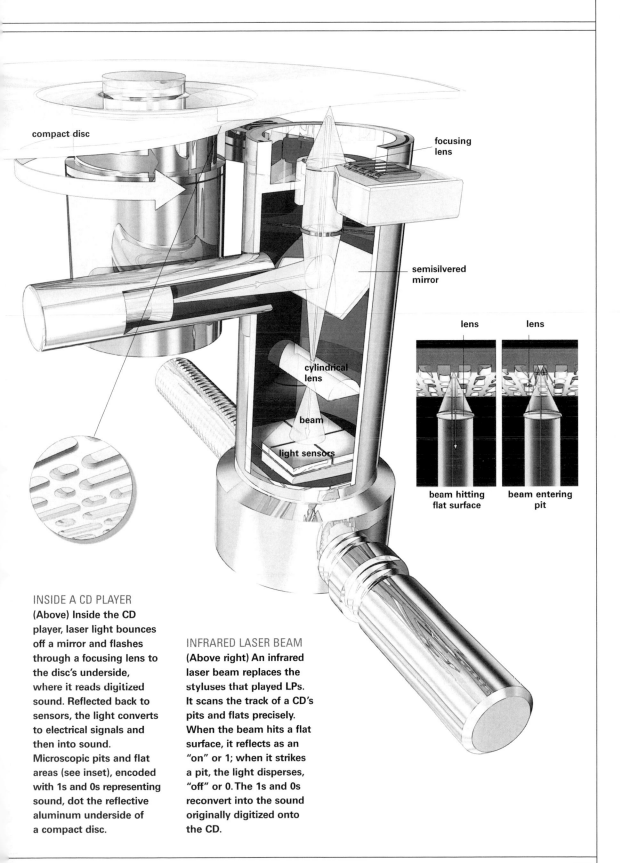

compact disc

focusing lens

semisilvered mirror

cylindrical lens

beam

light sensors

lens   lens

beam hitting   beam entering
flat surface   pit

INSIDE A CD PLAYER

(Above) Inside the CD player, laser light bounces off a mirror and flashes through a focusing lens to the disc's underside, where it reads digitized sound. Reflected back to sensors, the light converts to electrical signals and then into sound. Microscopic pits and flat areas (see inset), encoded with 1s and 0s representing sound, dot the reflective aluminum underside of a compact disc.

INFRARED LASER BEAM

(Above right) An infrared laser beam replaces the styluses that played LPs. It scans the track of a CD's pits and flats precisely. When the beam hits a flat surface, it reflects as an "on" or 1; when it strikes a pit, the light disperses, "off" or 0. The 1s and 0s reconvert into the sound originally digitized onto the CD.

TELEVISION

THE POINT MAY BE ARGUED, BUT TELEVISION WAS NOT DEVELOPED by one individual. Most of the credit goes to two inventors of the 1920s. Britain's John Logie Baird developed the picture tube and was the first to televise moving objects. Vladimir Kosma Zworykin, a Russian-born U.S. engineer, developed an electronic scanning device and an electronic camera called the iconoscope. As Zworykin conceived it, a picture could be transmitted between distant points by shining it into a mica disk covered with a mosaic made of photoelectric material. The disk would be scanned by a thin electronic beam to search for weak and intense emissions. Modern television transmission borrows from Zworykin's idea and from radio transmission, which uses radio-frequency carrier waves to transport information.

Essentially, a TV camera changes light from an image into a video signal by breaking the subject into an arrangement of 525 to 625 "lines of resolution," which are then scanned electronically, line by line. The sequence of pulses arrives at the TV set, which turns them back into scan form. The scan is reconstituted as an optical image on a screen that is coated with chemicals sensitive to the three primary colors of light: red, green, and blue.

Technology has transformed the traditional boxy television set into a sleek panel that screens extraordinarily vivid images. For one thing, the shift from analog to digital transmission meant an increase from 525 scan lines to more than 1,000 lines, as in high-definition digital television, or HDTV.

A cathode ray tube needs to be fairly deep to produce a wide picture. Flat panel sets, either plasma or LCD (liquid crystal display), provide a solution. Plasma displays work by lighting up minuscule fluorescent lights—red, green, and blue—to produce an image. Plasma, a key component in fluorescent light, is an electrically charged and conducting gas which has been referred to as a fourth state of matter. In the form of xenon and neon in a plasma TV set—where the gas is sandwiched between two glass plates in tiny cells—its atoms release light particles, photons, when excited.

LCD sets rely on liquid crystals (substances that flow like a liquid but retain some of a crystal's characteristics). Inside a screen, liquid crystal-filled cells—again red, green, and blue—get an electrical charge that shifts the crystal to a certain angle which, when illuminated by a light behind the screen, controls the images.

PICTURE TUBE
(Right) Guided by a magnetic field and fired out of three electron guns, electron beams that correspond to colors in a TV image strike millions of dots of fluorescent compound on the inside of the screen.

BIG SCREEN
(Above) A wide-screen HDTV plasma monitor showcases to advantage a blockbuster film at an electronics store in Niles, Illinois.

SEE ALSO

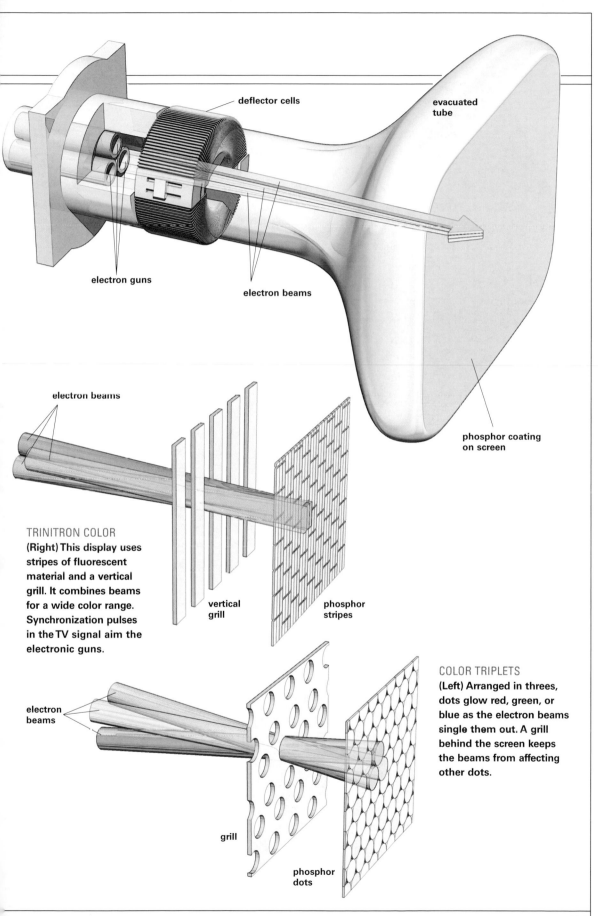

deflector cells

evacuated
tube

electron guns

electron beams

phosphor coating
on screen

electron beams

TRINITRON COLOR
(Right) This display uses
stripes of fluorescent
material and a vertical
grill. It combines beams
for a wide color range.
Synchronization pulses
in the TV signal aim the
electronic guns.

vertical
grill

phosphor
stripes

COLOR TRIPLETS
(Left) Arranged in threes,
dots glow red, green, or
blue as the electron beams
single them out. A grill
behind the screen keeps
the beams from affecting
other dots.

electron
beams

grill

phosphor
dots

VIDEO GAMES

WHEN PEOPLE OPERATE VIDEO GAMES, CHANCES ARE THEY DON'T know what creates the special effects. The machine may be a stand-alone console in an arcade, a compact handheld model complete with a microprocessing minicomputer inside, or a personal computer turned into a game center by binary imagery stored on a CD-ROM. Whatever its form, a video game is a collaboration of animators, 3-D graphic artists, audio programmers, interactive designers, software specialists, and electronics technicians.

Game concepts often begin with a design outline on a sheet of paper. An array of penciled connect-the-dots represents electric signals that stand for commands, screen action, background, sound, color, and movement of the images. In a process called rendering, video game programmers "draw" images on a screen with the help of 3-D graphics microprocessors or special software programs. Animation is achieved frame by frame by using computer programs that create and play back the artwork for editing. This process is an extension of the noncomputer technique of creating, say, animated cartoons by filming a series of images hand-painted or drawn on plastic "cels." Color and brightness are controlled by manipulating pixels (picture elements), the tiny bits of colored light that make up a video display. After images have been created, they are installed in a computer-generated background and stored as signals that can be resurrected and directed with a flick of the wrist.

The devices that enable video games are fairly complex. Game consoles are fairly bursting with microprocessors, memory chips, wireless controllers, and analog and digital audio-video outputs. Some devices, like those that let kids do digital microscopy, for example, are plugged into a PC that displays magnified images from a special digital microscope and camera. Games that are plugged into a television set use game cartridges or have controllers containing games and necessary hardware built in. Others are interactive, like one that has an electronic home plate, ball, and a plastic bat that allows the player to simulate hitting a baseball. The batter faces the TV screen standing above the electronic home plate, which emits an infrared beam, and swings at a simulated ball. An electronic system in the bat sends the swing action to the home plate, and the result is displayed on the screen.

TARGET PRACTICE
(Right) Through virtual shooting at a video game arcade, marksmen and markswomen can hone their target skills with a realistic display and a make-believe gun that fires electronic "bullets."

SEE ALSO

PORTABLE FUN
Handheld video games complete with sound effects and liquid-crystal color graphics can be played on batteries just about anywhere, and what's under their hoods often mimic some of what's inside a home PC. Palm-size, button-operated portables let one play games anywhere. Some are larger, others even smaller, and manufacturers seek to create more sophisticated devices year by year. Peripherals can connect the game to another console, allowing four players to play each other, using their own game boxes, or connecting the game to the Internet via a mobile phone to download other games, trade data, and play with others in real time.

CAMERAS

ARISTOTLE AND LEONARDO DA VINCI WERE WELL ACQUAINTED with the concept behind the camera obscura: Light rays from an external object enter a darkened chamber through a tiny hole in one wall, converge and cross, and project an inverted image of the outside scene on the opposite wall. Make the room smaller and add a lens, mirrors, prism, shutter, and film, and you have a no-frills camera.

A standard (nondigital) camera is basically a lightproof container that focuses, through a lens, light from a scene onto light-sensitive film inside. When a camera's shutter release button is pressed, the shutter opens, and light momentarily bathes the film. Too much light makes overexposed, washed-out images; too little, underexposed, dark ones. To regulate the amount of light that strikes the film, the shutter speed is often adjustable; in modern cameras the normal range is from several seconds to 1/1,000 of a second. Camera apertures, or "eyes," can also be adjusted to change a picture's sharpness and contrast.

Instead of using film to capture the image, a digital camera uses an arrangement of charge-coupled devices—light-sensitive semiconductors that can store electrical charges. These digitals sensors transform the light from the scene snapped into an analog signal that is then converted into a digital version. An electronic filtering system regulates color and other picture components, while another system reduces the picture. The image is finally sent to a temporary storage area, then onto a memory card. One obvious advantage of a digital camera is its storage capacity: hundreds to a card, as compared with up to 36 on a roll of film.

CLOSE UP

(Below) A photographer trains his camera and flashes on a treetop frog in the Borneo rain forest. Exchangeable lenses offer the same camera a variety of distance-focusing and magnification capabilities. So-called macro lenses, along with special filters, allow close-up work.

SEE ALSO

Photocopiers · 160
Lenses · 174
Telescopes · 180

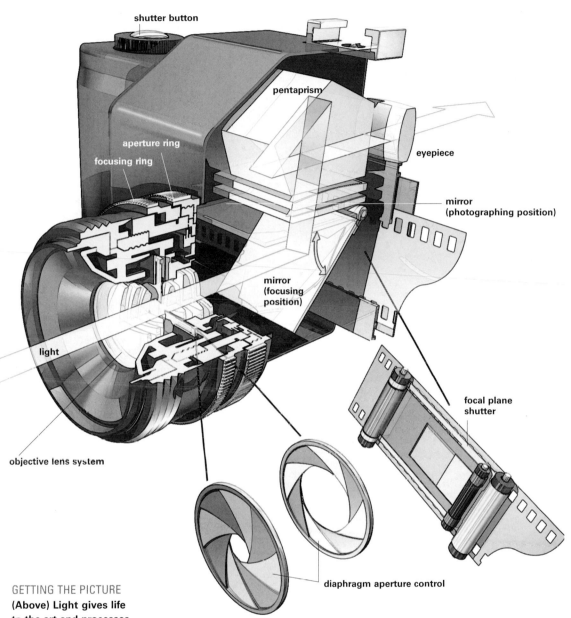

shutter button

pentaprism

aperture ring

focusing ring

eyepiece

mirror
(photographing position)

mirror
(focusing
position)

light

focal plane
shutter

objective lens system

diaphragm aperture control

GETTING THE PICTURE

(Above) Light gives life to the art and processes of photography. In a camera, a mirror and a prism correct an inverted, reflected-light scene, and a viewfinder lets a photographer see it through the camera lens, which is actually a system of optical lenses. Adjusting the distance between the lens and the film brings an object into clear focus. When the photographer presses the shutter button, a spring-activated device opens and closes the aperture. Light is kept out of the camera body except during exposure. With the aid of a diaphragm—a fixed or adjustable component forming an opening—the shutter allows the correct amount of light in through the lens. As the light strikes the film at the rear of the camera, it leaves an imprint of the scene.

MOTION PICTURES

I N 1872, LELAND STANFORD BET A FRIEND $25,000 THAT A RUNNING horse had all four feet off the ground at some time in its stride. When this railroad builder and former California governor asked photographer Eadweard Muybridge to help him prove it, Muybridge hit on a novel solution. He lined a racecourse with cameras, attached a string to each shutter, and stretched the strings across the track. As a horse galloped along, it broke the strings, releasing the shutters. The plan won Stanford his bet by portraying a series of still shots of the stride of a galloping horse—something that artists had never done accurately.

Modern motion pictures are technology-driven and awash in special effects, but they are nonetheless generally dependent on film, lenses, and single pictures. A conventional movie camera captures movement on a strip of film that is drawn past the lens aperture and stopped for a fraction of a second at timed intervals. When the film stops, the shutter opens quickly and exposes one frame, or picture; when the shutter closes, the film advances to the next frame. This sequence repeats at the rate of 24 exposures a second. Projected on a screen, a movie is an illusion of motion made possible by the fact that the human eye briefly holds onto each one of the single images until the next one replaces it.

Movie film still has the best quality, at least for the moment, but in the view of many filmmakers, digital-camera moviemaking will eventually win out. Film editing and special effects are already done digitally, filmmaking schools teach digital techniques, and independent filmmakers, without the Hollywood budgets, are producing digital movies and showing them over the Internet and via cable TV.

MOVIE INDUSTRY

(Above) George Lucas's *Star Wars* films made movie history with their advances in camera work, film processing, and special effects.

MOVING PICTURES

(Right) As film passes through the projector, the rotating shutter opens for a fraction of a second, projects an image, then closes and cuts off light through the film. In a motor-driven movie projector, a claw moves a perforated reel of film past a bright lamp and a mirror-reflector behind the film. A shutter blocks out light while film moves frame by frame, from reel to reel.

SEE ALSO

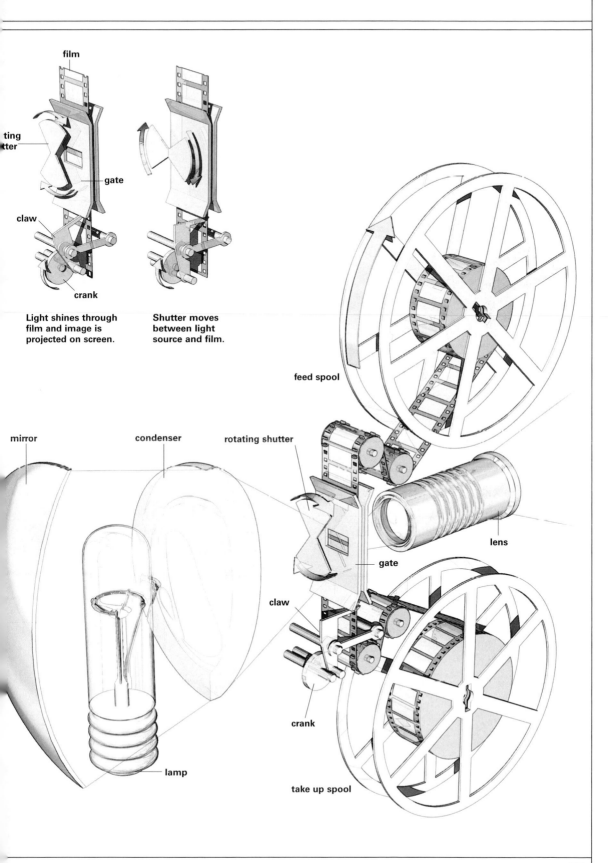

film

ting
tter

gate

claw

crank

Light shines through film and image is projected on screen.

Shutter moves between light source and film.

feed spool

mirror

condenser

rotating shutter

lens

gate

claw

crank

lamp

take up spool

ROLLER COASTERS

T HEY HAVE INTIMIDATING NAMES: MILLENNIUM FORCE, Dragon Kahn, Cyclone, Beast, Mean Streak, and Run with the Bulls. They can move you hundreds of feet up a ramp and propel you to the bottom at a 65-degree angle in a stomach-churning, bone-shaking plunge that takes only seconds. Roller coasters are the ultimate thrill ride, a mode of g-force transportation in open cars reaching one hundred miles an hour, and they are designed for one purpose only—to give riders an adrenaline rush.

Roller coaster car-trains slowly drag you up the tracks of a wooden or steel scaffold in anticipation and apprehension, then they crest the first peak of the rail and carry you with them as they seem to freefall around pretzel-like twists and turns, leaving you in pure fright. These amusement vehicles have their origin in sledlike, hollowed-out blocks of ice that raced down ice slides in 15th-century Russia. Then, as now, they fulfilled what appears to be an age-old lust for heart-stopping, breath-holding sensation. The faces on roller-coaster thrill-seekers say it all—which is that rides that twist, jerk, shake, and otherwise stress the human body may be scary to some, but for others they offer a nerve-tingling escape from immobile boredom.

The first roller coaster built in the United States was conceived in 1884 by LaMarcus Thompson. Thompson called his invention the Gravity Pleasure Switchback Railway. Operated at Coney Island, New York, its cars were hauled manually to the top of a 45-foot incline before being released on a six-mile-an-hour ride. The cars were hauled up empty, and passengers climbed up a long flight of stairs to get into the cars for their big dose of downhill excitement. Thompson charged five cents for a ride and made hundreds of dollars a day, the ride was so popular.

Today's roller coasters are still basically gravity machines. As such, they follow Isaac Newton's first law of motion, which dictates that a body at rest stays at rest, while a body in motion keeps moving until someone—or something, like the force of friction—applies the brakes.

Drawn to the top of an incline by a motor and chain, a roller coaster rolls along on the sheer momentum from the first descent. The ride ends when a combination of friction, wind resistance, and braking slows the cars down to a speed of zero.

WHIRLWIND RIDE

(Right) According to the International Association of Amusement Parks, a rider's chance of receiving a serious injury on a roller coaster is one in four million; the chance of a fatal injury on a roller coaster is one in three hundred million.

SEE ALSO

ENGINEERING THRILLS

Roller coaster engineers will do anything to extend the thrills. IIn an effort to create the illusion of higher speed, even though the roller coaster is slowing down as it nears its final resting point, ride engineers have added sharper curves at the end of a coaster track to take advantage of residual energy. Linear induction motors are being used to achieve faster speeds and to help the roller coaster maintain speed as long as possible. One of the world's fastest roller coaster is a Japanese machine that hit 107 miles an hour on its maiden run. Its spectacular track includes a close-to-vertical drop of 170 feet, which the car takes before it goes hurtling around an equally thrilling bend.

FIREWORKS

FIREWORKS HAVE ELICITED OOHS AND AAHS FROM DELIGHTED observers ever since the ancient Chinese gave the world an early form of rockets, Roman candles, and fiery pinwheels. They may have invented pyrotechnics for use as weapons or to scare away demons, but not much time went by before the Chinese realized that fireworks would play very well as outdoor entertainment.

Fireworks technology of the tenth century was not so different from that of today. It relied on gunpowder mixed with various chemicals to provide color, and it included metal shavings that would create the sparkle effect. Fireworks now are generally tube-shaped or spherical paper containers packed with explosive powder and a time-delay fuse. The containers also hold strategically placed packets of metallic salts. Lithium or strontium produces red, barium nitrates make green, copper compounds result in blue, sodium creates yellow, charcoal and steel produce sparkling gold, and titanium makes white. Powdered iron, aluminum, or carbon produce the sparks and other special effects.

A pyrotechnist places a fireworks shell in a mortar or a special gun and lights the shell's main fuse. The lift charge ignites and propels the shell, then the time-delay fuse ignites the chemicals inside, creating patterns and shapes depending on their placement.

Not all fireworks are launched, however. Some of the more complicated ones stay on the ground, or are attached to poles or trees and spin about on wheels, their motion produced by the recoil as the fire escapes from cases. These floral fountains send a shower of sparks upward. The familiar handheld sparkler likewise sprays its fire up and out, as pyrotechnic materials coating the base stick burn from the top down.

BRILLIANT BLOOMS
(Right) Floral fire paints the sky as a time-delay fuse detonates fireworks in a carefully timed, multicolored light show.

FIREPOWER
(Below) Rolled paper and glue hold a fireworks shell together. When ignited, the leader fuse burns down to a lift charge of black powder and propels the shell skyward. Inside, a time-delay fuse causes the color-producing chemicals to ignite and scatter; the fuse can also set off the noisemaking salutes. For full effect, pyrotechnists time shells to explode at the highest point of their trajectory.

SEE ALSO

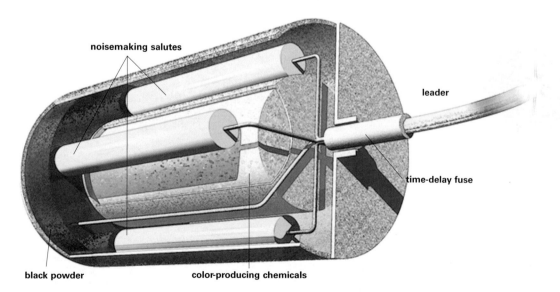

noisemaking salutes

leader

time-delay fuse

black powder

color-producing chemicals

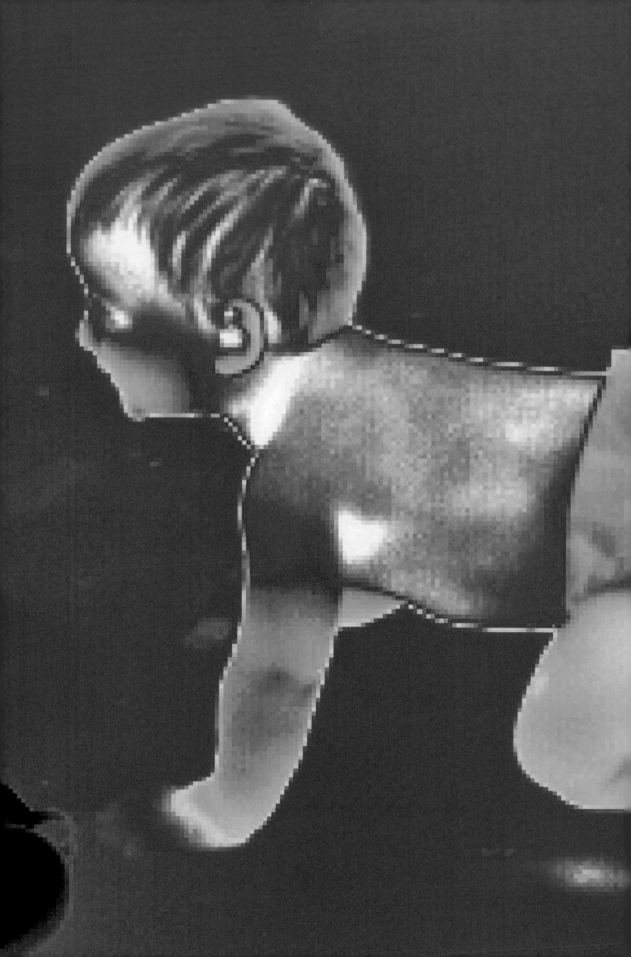

CHAPTER 8

HEALTH & MEDICINE

Health care today is a work of both science and art. Sophisticated imaging systems see structures that eyes alone can never see; ultrasound hears a rush of blood that ears cannot. Satellites beam information to doctors who are far removed from the surgical scene, and robots wield surgeons' tools. Medicines are designed by computers and delivered by patches, sprays, and implants. Surgery is performed through tiny incisions. Reengineered genes are piggybacked onto viruses and steered through a patient's body to repair a defect or replenish a deficiency. Radiation pellets and rays treat cancers. Gerontologists are even exploring ways to tinker with the genetic clock, hoping to slow it down a bit. If they succeed, and if we can eradicate major diseases, we will live out our lives with perhaps a score or more years added, and we will do so in good health.

A diagnostic thermographic image shows the baby's cold feet in mauve and its hotter forehead in yellow.

ENDOSCOPY

I N THE CLASSIC SCIENCE-FICTION MOVIE, *FANTASTIC VOYAGE,* A GROUP of scientists and their submarine are miniaturized and injected into a patient, to get a first-hand look at organ systems and the interaction of invader antigens and defending antibodies. Endoscopy affords a somewhat similar view. With flexible fiber-optic tubes inserted through natural openings such as the nose, mouth, anus, urethra, and vagina, doctors can examine a person's internal organs and structures. One form of endoscopy—a colonoscopy—lets a diagnostician search the large intestine for bowel diseases and bleeding. Another approach requires a small incision in the abdominal wall, into which is inserted a lighted instrument called a laparoscope. Cystoscopy explores the urinary bladder to diagnose bladder stones or cancer. Rhinoscopy examines the nasal passages and the rear of the throat. Special instruments may be attached to such tubes, allowing surgeons to remove lesions and collect samples.

Technological advances allow surgeries to be performed with minimal invasion into the body, thanks to a flexible medical instrument called the endoscope. Inserted through natural openings or small incisions made in the body, this device lights its own way with a bundle of optical fibers built into it. Peering through an eyepiece, a physician can manipulate controls that send signals through a channel and maneuver surgical instruments, to perform a procedure or take a tissue sample. With the appropriate fittings, endoscopes can also cauterize bleeding vessels and suction out tissue debris.

Virtual endoscopy, a blending of CT and MRI scans with high-performance computing, offers a noninvasive way to get a direct look. Scans can provide simulated visualizations of specific organs in 3-D animation. A virtual endoscopist sits at a workstation and views the inner anatomy by manipulating a computer mouse. Diagnosticians change the angle of view, and scale or shift immediately to new views.

Increasingly, surgeons use the technique of endoscopy to remove ovarian cysts and herniated lumbar disks, do hysterectomies and bowel resections, or repair hernias and torn knee ligaments. A gall bladder operation once required a five- to eight-inch incision and a recovery period of up to two months. Now doctors make three tiny incisions in the abdomen and through them insert surgical instruments and a small video camera. They separate the gall bladder from the liver and other nearby structures and remove it through one of the incisions. The procedure is done on an outpatient basis and requires a two-week recovery.

FETAL SURGERY
(Right) Minimally invasive surgical techniques mean that doctors can treat birth defects while a fetus is still in the uterus.

ENDOSCOPE
(Below) This flexible instrument lights its own way and peers into the body with a bundle of optical fibers.

SEE ALSO

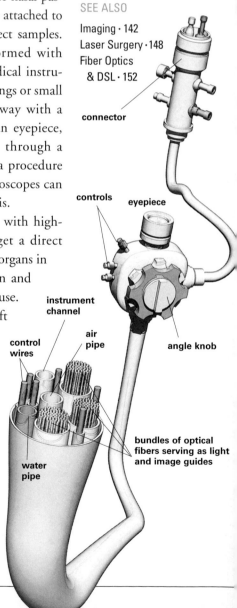

connector

controls

eyepiece

instrument channel

control wires

air pipe

angle knob

water pipe

bundles of optical fibers serving as light and image guides

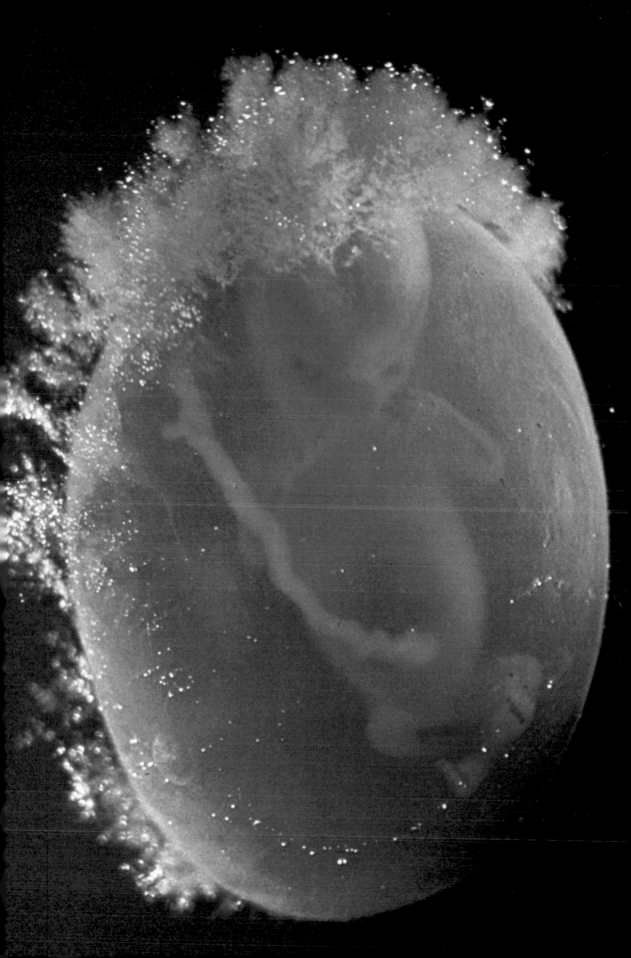

IMPLANTS

MODERN TECHNOLOGY DESERVES ENORMOUS CREDIT FOR giving patients artificial heart valves and prosthetic limbs, implantable cardiac defibrillators, titanium and ceramic hip joints, and numerous other spare parts. Indeed, according to the U.S. Food and Drug Administration, more than 20,000 firms worldwide produce more than 80,000 brands and models of medical devices for America's market.

An implanted pacemaker with a tiny transmitter can automatically send data about the patient's heart to the physician. A wristwatch-like device for diabetics monitors the wearer's glucose level every 20 minutes; an alarm sounds when it gets dangerously high. A capsule-camera, once swallowed, can take pictures of internal bleeding and other abnormalities as it travels through the small intestine.

Suitable materials are the key to successful implants. Aluminum compounds are used in dental and orthopedic prostheses because they interact minimally with surrounding tissue and experience little friction or wear. Synthetic polymers that resist water and oxidation and that can also insulate and lubricate are used to protect electrical implants, such as cardiac pacemakers. A number of porous materials and absorbable composites allow for natural tissue growth in and around some implanted devices; special coatings on metal implants reduce corrosion. A coating of genetically engineered cells may one day prevent an implant from being rejected or choked by the scar tissue made when a body is invaded.

Prosthetics have also evolved. An amputated arm, for example, can be replaced by a prosthesis powered by a harness and cable attached to the residual limb, or one that uses an external powered device. New imaging devices have ensured near-perfect fits for artificial limbs, and myoelectric devices have been developed that can shift the electrical impulses of a person's own muscle contractions to a prosthetic limb, allowing for a relatively natural range of movement. Computerized prosthetics that regulate gait can move joints by controlling a hydraulic and motor-driven system. Implantable electrodes are used to stimulate muscles in spinal cord injury patients. In the future, according to some, brain waves may even power prostheses.

BONE IMPLANT

(Right) Metal pins inserted into an ankle stabilize and help repair a fracture. The range of such metallic medical hardware is prodigious and versatile. Nails secure broken bones; hip screws, spinal fixation plates, leg-lengthening apparatuses, and locking bolts take care of other structures.

SEE ALSO

IMPLANTED HEARING

Common hearing aids use battery-powered microphones, amplifiers, and earphones. The cochlear implant, far different, can often benefit those who cannot hear even with a hearing aid. It is placed in the ear's cochlea, a snail-shaped section of the inner ear. It consists of an internal and external coil, a microphone, a signal processor that converts the sound into electrical signals, and electrodes that send electrical signals into the cochlea. One coil is surgically implanted behind and above the ear, with electrodes connected into the cochlea. The external coil sits over the site of the internal one and transmits electrical impulses through the skin, thus stimulating the auditory nerve.

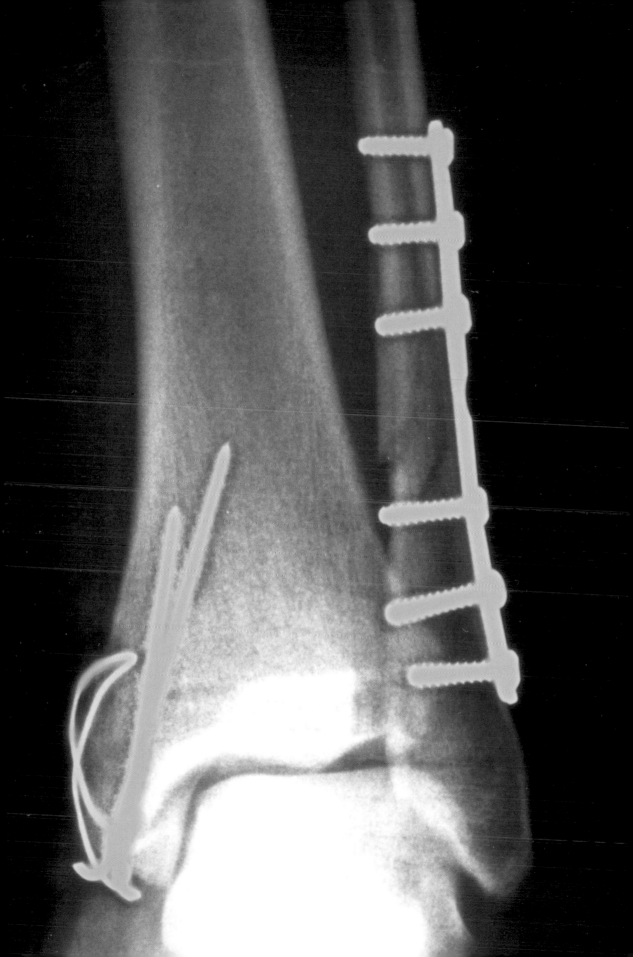

IMAGING

I F DIAGNOSTIC EQUIPMENT HAS A VENERABLE GRANDFATHER, IT IS THE x-ray machine. Still the quickest way to take off our skins and pose in our bones, as someone once said, it works through the use of controlled beams of highly energetic electromagnetic rays, discovered by Wilhelm Conrad Roentgen, a German physicist, in 1895.

X-rays pass through flesh and thick paper but are stopped and reflected by bones and metal. When the rays then strike a photographic plate, they result in an image of bones in white. Various shadows and shadings help identify anomalies, such as breaks or fractures.

Diagnosticians need more refined tools, though, because x-rays project only two dimensions and can miss many structures and abnormalities. Computerized axial tomography, called the CT scan, often pronounced *cat scan,* has been hailed as the greatest advance in radiology since Roentgen's discovery. It emerged in 1972 largely through the research of Dr. Allan MacLeod Cormack, who shared a Nobel Prize for his work. Linking x-ray and digital technology, a CT scan shows the body in cross-sections, from which 3-D images are constructed. Differences in density between normal and abnormal tissues are revealed, as well as bone details and the location of tumors and other signs of disease.

Magnetic resonance imaging (MRI) also produces cross-sectional pictures. While a patient lies in a tunnel-like chamber, surrounded by electromagnets creating an intense field, the MRI unit generates and reads the radio signals that return from hydrogen atoms in the water molecules of body tissues. A computer converts the signals into images of soft tissue, such as that in the brain and spinal cord.

A version called fMRI, for functional MRI, is being used by neurologists to look at nerve cell activity and brain function, evaluate patients with Alzheimer's disease and schizophrenia, and assess disease-related changes in cerebral blood circulation. Extremely rapid imaging techniques provide pictures in a few seconds and, in fact, a development called echoplanar imaging (EPI) allows image acquisition in a fraction of a second.

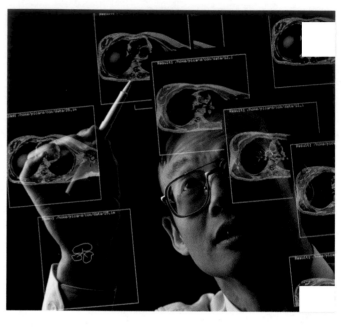

HEART SCANS

(Above) Digital scans of a human heart can help physicians detect atherosclerotic plaque in arteries and analyze valve function and a variety of conditions that damage or overtax the heart muscle.

SEE ALSO

Endoscopy · 138
Fiber Optics & DSL · 152

MAGNETIC RESONANCE

(Below) Electromagnetic coils in an MRI unit scan a body with radio waves, exciting hydrogen atoms in bodily tissues. The imaging device then reads signals returned by the atoms to create cross-sectional, 3-D views. Low-powered magnets, called shim coils, control the main magnetic field to vary field strength, setting up gradients in different planes. The coils marked X read from left to right; Y, from front to back; and Z, from head to toe. In this way, each portion of the body can be identified with magnetic coordinates and rendered into a computerized sectional image.

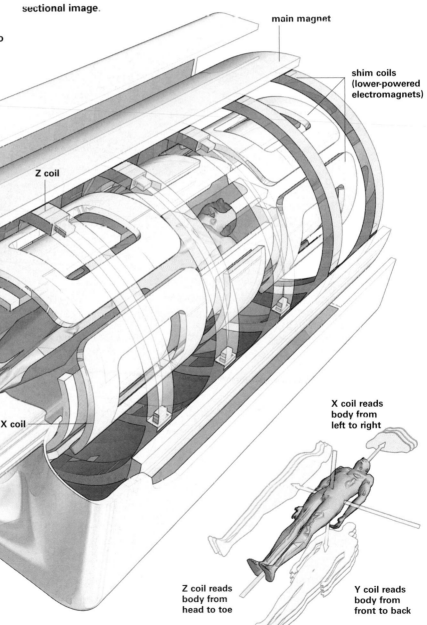

main magnet

shim coils
(lower-powered electromagnets)

Z coil

Y coil

X coil

X coil reads
body from
left to right

Z coil reads
body from
head to toe

Y coil reads
body from
front to back

MONITORS

I T IS QUITE IMPOSSIBLE FOR ANY MAN TO GAIN INFORMATION respecting acute disease, unless he watch its progress," said the eminent 19th-century English physician Richard Bright, "the lapse of eight-and-forty hours will so change the face of disease." One of the foundations upon which medicine is built, in fact, is observation. Medical monitoring systems hum and beep in hospitals' surgical suites, in emergency rooms and in patient rooms and in the intensive care unit, where minute-by-minute observations can make the difference between life and death.

Much of today's monitoring instrumentation is grounded in some of the old standbys. The electrocardiograph still traces the heart's electric current. Electrodes pasted to the chest, legs, and arms help chart ventricular contractions and pinpoint anomalies that indicate disease and disorder. The electroencephalograph records electrical activity in the brain, looking for telltale peaks and dips that signal tumors. The pulse oximeter, attached to a finger or earlobe, searches for inadequate amounts of oxygen in the blood by monitoring the percentage of hemoglobin, the complex compound that sends oxygen to other cells.

Medical monitoring instrumentation has become increasingly smarter, though—computerized and transformed, either as stand-alone apparatus or as part of a network linked to a central nursing station where the data can be observed.

For all its simplicity, the oximeter is a piece of electronic ingenuity. Linked to a computerized unit that displays the amount of oxygen-saturated hemoglobin in the blood, it also provides an audible pulse-beat signal, calculates the heart rate, and monitors the rate of blood flow. It does its job by analyzing the wavelengths of a light source, since light is absorbed in differing amounts by hemoglobin, depending on the saturation of oxygen.

Microprocessors, at the heart of the computerized systems, do away with yards of wiring, allowing for more precise measurements of temperature, blood pressure, pulse rate, and oxygen levels in the blood. In the past, instruments that monitored such essentials were elaborate affairs, but today a single computer chip can be programmed to take many of these important measurements.

Other monitoring devices are portable, such as the battery-operated telemetry monitor that checks heart rate and other activity. Carried about in a pocket or holster like a cell phone, it is connected to chest patches; the heart's electrical activity is then transmitted wirelessly, by radio waves, to a monitor screen in the patient's room or at a nursing station. Aside from giving the patient more mobility during tests, wireless medical telemetry can reduce health care costs because it allows several patients to be remotely monitored simultaneously.

MOBILE CARE
(Right) With an array of monitoring and life-support equipment close at hand, emergency medical technicians tend to a patient in their speeding ambulance.

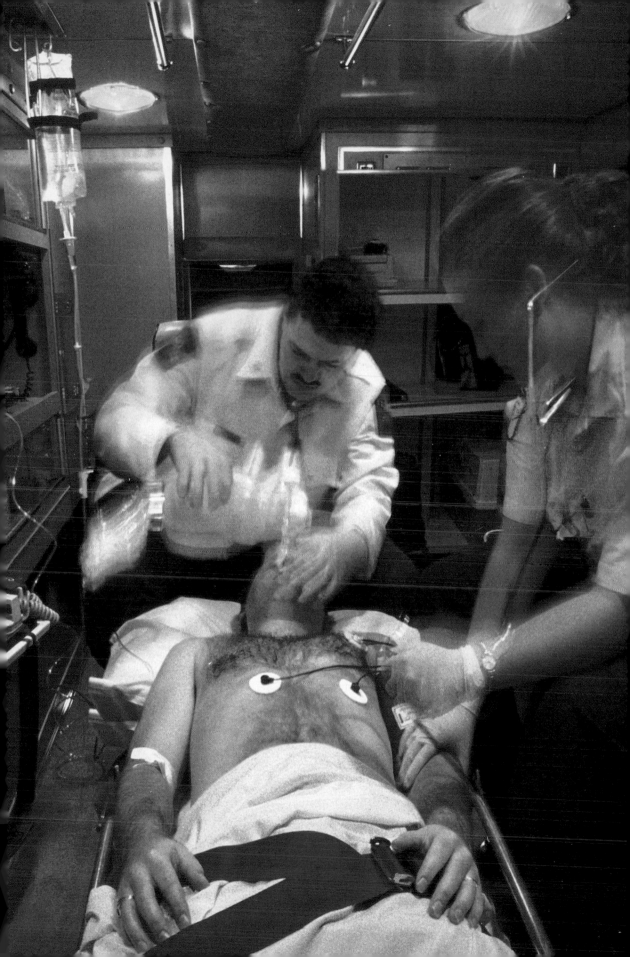

DRUGS

CANADIAN PHYSICIAN AND EDUCATOR SIR WILLIAM OSLER had his finger on the pulse of hypochondriacs and quick-fixers when he said that "the great feature which distinguishes man from other animals" is the desire to take medicine. In just one year, pharmacists in the United States fill some two billion prescriptions. The country's drug companies pour 24 billion dollars annually into research and development, spending a total of 500 million dollars, on average, for each new drug that reaches the market.

Traditionally, creating a drug involved adding compounds one at a time to cell cultures and enzymes, then laboriously testing those to see if they needed to be modified. Computers have eliminated some of the drudgery by simulating, for example, protein receptor sites on a virus and then modeling a molecule that might fit the receptor.

Another approach to drug design involves inducing animals and crop plants to make therapeutic proteins and antibodies. One technique coaxes goats and cows to secrete a selected protein in their milk, and another gets maize plants to produce disease-fighting antibodies in the nutritive tissues of their seeds.

Taking drugs once meant swallowing pills or getting an injection, but these techniques do not always break down a drug or send it into the bloodstream effectively. Newer delivery methods can compensate for the shortcomings: Drug-impregnated, foil-backed skin patches bypass the digestive system and let medication be drawn into the skin via a tiny electric current. Nasal sprays take advantage of the nose's permeable mucous membrane, a portal into the circulatory system. Biodegradable drug implants, insulin pumps, and time-release medications all hold promise.

PILL PANTRY

(Right) A pharmacist selects a patient's pills. "For every ill a pill," said British writer and social critic Malcolm Muggeridge.

FINDING THE KEY

(Below) A lock-and-key mechanism relying on cell receptors helps drugs work within body cells. A drug (below, left) mimics and reinforces natural messenger molecules that the body produces to call up disease-fighting white blood cells. Another drug (below, right) blocks receptor sites to keep out molecular messages and allow the cell to function normally. Pharmacologists can design drugs with molecules that match protein receptor sites on viruses or disease-related enzymes, fooling the sites into accepting the new drugs.

SEE ALSO

message

**Some drugs reinforce
natural substances**

**Some drugs block
natural substances**

receptor site ——

natural body —— substance

reinforcing —— drug

drug blocking —— receptor sites

LASER SURGERY

ASERS CARRY THOUSANDS OF TELEPHONE CALLS SIMULTANEOUSLY over fiber optic cables. They play CDs, slice steel beams, guide missiles, measure distances, fashion suits and semiconductor chips, and bore holes in diamonds in a fraction of a second. Lasers also help heal. Medical lasers can vaporize brain tumors, spot-weld tissue grafts, fragment kidney stones, cauterize the lining of a uterus to stop prolonged bleeding, and clear blocked fallopian tubes. In addition, cosmetic lasers remove unwanted hair, obliterate tattoos and birthmarks, smooth facial wrinkles, and activate tooth-bleaching solutions.

Used in surgery, a laser's work may be called bloodless or knifeless; it seals blood vessels as it cuts, and it sterilizes at the same time. Its precision makes it ideal for microsurgical procedures, such as repairing detached retinas. Laser stands for "light amplification by stimulated emission of radiation," but its light differs from that of the sun or a bulb, which radiates in every direction. Concentrated and narrow, laser beams move in the same direction. Moreover, they can focus intense heat into small spots: At a width tinier than a pinpoint, a laser can generate a temperature of 10,000°F. And because human tissue is about 80 percent water, a laser can vaporize diseased cells in its target zone.

Wavelength and power are determined by a laser's makeup. Argon gas is used as the laser medium for certain eye surgeries, such as the repair of retinal holes and tears, because the laser's blue-green light passes easily through the eye's cornea, lens, and fluid before it reaches the retina. The carbon dioxide laser used in gynecology has a beam that is absorbed by substances containing water. A "biostimulation" laser (also known as a cold laser or a soft laser) provides low-level therapy, such as that required in acupuncture, used as an adjunctive device for temporary relief of pain.

HEALING LIGHT

(Right) Aiming the intense beam of a head-mounted laser, a surgeon at Johns Hopkins University's Wilmer Eye Institute repairs a retina. Ophthalmologic lasers can also destroy tumors and arrest abnormal growth of blood vessels.

SEE ALSO

LASER ERASER

Able to cut and dissolve tissue, medical lasers differ in their use and power according to their makeup. Surgeons can use lasers to remove tattoos, because the short pulses of intense light beams pass through the skin and are absorbed by the tattoo pigments. When that happens, the pigments break down into particles small enough for the body to absorb them and flush them away. Black pigment is the easiest to remove. Blistering and scabbing may follow, but in general the laser does not inflict damage to the skin itself. Laser treatments like this and others rarely require anesthesia. In general swift and focused, they are usually performed on an outpatient basis.

CHAPTER 9

INFORMATION

W E COMMUNICATE INFORMATION IN many ways, passing it along through our speech and writing, symbols and signs. We also use visual images, gestures, facial expressions, and bodily attitude as we exchange and share facts and notions. Animals have many ways of communicating with one another, and plants may have a system of communication as well, but as far as we know, only we humans have developed the ability to reconstruct and vastly improve our ways of communicating. To do so, we tame and bridle such natural forces as electricity, radio waves, light, and sound, using them alone or in combination. From the cellular telephone to e-mail, from telecommunication devices for the deaf to wearable computers, science and technology have left an indelible imprint on how we get our messages across.

This tiny Intel Pentium chip is capable of controlling a computer's behavior with its millions of transistors and circuitry.

FIBER OPTICS & DSL

TELECOMMUNICATION—COMMUNICATION AT A DISTANCE USING telephones or radios, for example—relies on electromagnetic waves. Members of this large energy family include radio and light waves, x-rays, microwaves, and infrared and ultraviolet rays. Called electromagnetic because they consist of electric and magnetic fields (which vibrate at right angles to each other), they are high-velocity energy transmitters that move at the speed of light, 186,000 miles a second. Their signals and impulses are capable of carrying sounds, words, images, and data. The higher their frequency, the more energy they have; gamma and x-rays are at the high end; microwaves and radio waves are at the low end.

Radio waves exist naturally as radiant energy from the sun, but when they are generated by electricity in an antenna tower, they transport sound and images. In the form of radar, radio waves help detect distant objects and determine their positions; radar equipment measures the time the waves take to travel to an object, reflect off it, and return. The visible light segment of the electromagnetic field has also been used to communicate: In laser form it beams through fine strands of pure glass, its digitally coded pulses translated into thousands of telephone conversations that can be handled simultaneously.

Sound also comes in waves, but they are pressure waves, not electromagnetic ones. Best defined as a measurable, mechanical disturbance, sound is amenable to translation into electricity. This is a valuable quality when transmitting voices, whose words, whether flowery or vernacular, must still be reduced to the same workmanlike electrical signals before they can be sent coursing over conventional copper wire or through tubes of glass.

But all these data-carrying signals can sometimes get in the way of one another, as is the case when one tries to make a phone call while someone else is accessing the Internet on the same line. Enter broadband, a special telephone line that now enables one to make a phone call and work on the Net simultaneously and on one line. Known also as DSL, for digital subscriber line, a broadband connection lets one surf the Net at a much faster speed than is possible over a regular phone circuit. To accomplish this, it splits the signals into two channels, one for voice and the other for the high-speed data. Bandwith, which indicates how much data can be transmitted through a connection, is the key here, and is generally measured at bits per second (bps). A conventional computer modem— an electronic device that converts digital data into analog signals to be sent over phone lines—can move data at more than 57,000 bps, but because a DSL modem squeezes more out of a phone line, the speed at which information is exchanged is greatly increased.

SHEATHED
OPTICAL FIBERS
(Right, below) Wrapped around a steel strengthening wire, these fibers transmit digital data in the form of laser pulses. Each fiber's core and cladding are made of silicon glass and semiconducting materials, and the plastic sheathing prevents stray light from escaping to other fibers. Narrow-core fibers (far right) carry clear signals over long distances. Lower-cost wide-core fibers used for shorter distances allow the signal to spread out and blur, limiting data's transmission rate. They are therefore used for local distribution. Optical fibers such as these carry information around the world at lightning speed for the purpose of communication, but also make noninvasive medical imaging possible.

FIBER OPTIC CABLES
(Right) Telephone conversations pulse digitally through cables like these. The glass strands carry tens of thousands of calls at the same time.

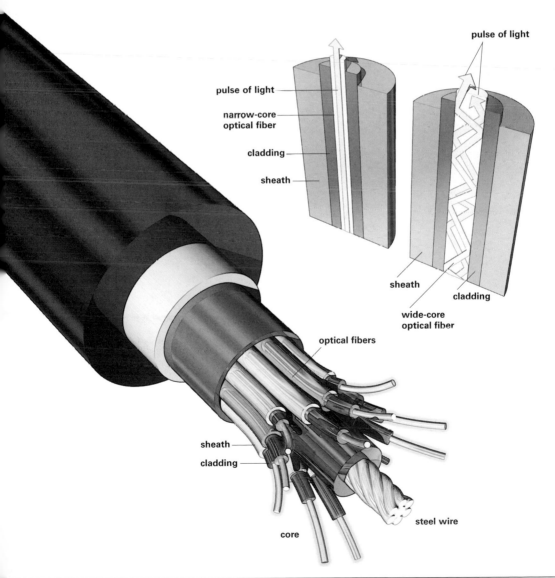

pulse of light

pulse of light

narrow-core optical fiber

cladding

sheath

sheath

cladding

wide-core optical fiber

optical fibers

sheath

cladding

steel wire

core

CELL PHONES

T HE WEDDING OF THE TELEPHONE AND THE RADIO WAS PERHAPS the most significant development in the history of telephony. Roaming from room to room with a cordless phone, or snapping open a compact cellular model while on a stroll in a park, a phone user now has an expansive degree of mobility.

Cellular phones transmit over radio waves via antennas located at base stations. Antennas are connected by phone lines to exchanges, which link cell phone users to one another or connect them to parties using conventional phones. Some cell phones signal through a satellite system, and one day global systems will enable callers to reach out to all corners of the world. Another generation of cell phones will be able to instruct a global positioning system satellite—alerted by a 911-like message—to beam a signal to the receiver, thereby determining the exact location of, say, a driver in distress.

At the center of it all is wireless technology, the same through-the-air transmission system that uses radio frequencies and other types of wave phenomena to make robot rovers scour the surface of Mars, send remote commands to control a TV set, open garage doors, power pocket computers, and let people talk to one another over two-way radios. While copper wire and fiber optics still handle many of our communications systems, the dream of a wireless world comes closer as people rely on palm-size PDAs, personal digital assistants that manage information such as appointments and addresses and can be attached to a wireless modem to surf Web sites and send and receive e-mail.

WALK AND TALK
(Right) A familiar sight all over the world, a cell phone to the ear has become an important means of communication.

CELLULAR NET
(Below) In cellular phone systems, signals travel via antennas at base stations located in cell-like areas or dishes oriented toward satellites. The radio frequencies in a cell group differ from one another. As a person moves from cell to cell, the call automatically switches to the appropriate frequency.

SEE ALSO

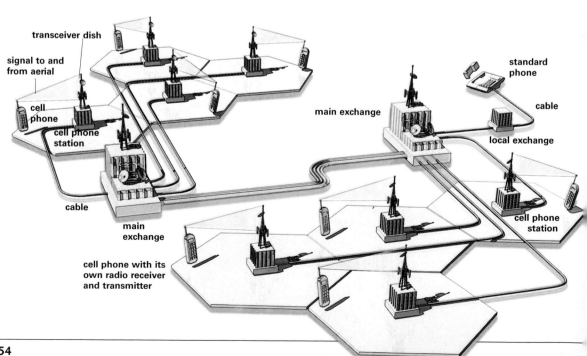

transceiver dish

signal to and from aerial

cell phone

cell phone station

cable

main exchange

cell phone with its own radio receiver and transmitter

main exchange

standard phone

cable

local exchange

cell phone station

SATELLITE COMMUNICATIONS

S PUTNIK I, THE FIRST ARTIFICIAL OBJECT TO ORBIT THE EARTH, was launched by the former Soviet Union in 1957. Today, half a century later, communication by radio, telephone, and television is unthinkable without these miniature moons. Carried into orbit by rockets or space shuttles, satellites transmit TV, news, and phone calls, track weather systems and monitor soil moisture, map Earth's features, broadcast navigational signals, help plan battlefield strategy, even carry up-to-the-moment financial information to Wall Street traders and allow us to swipe our credit cards through machines at gas stations.

Satellites can range over the planet, but the so-called geostationary orbiters used for communications and weather observations are synchronized with the Earth's rotation. They "park" above the same point on Earth's surface, a position that allows for uninterrupted contact between ground stations in a satellite's line of sight. Three relay satellites uniformly spaced at such a height can easily cover the entire surface of the planet, receiving television and hundreds of thousands of telephone messages from one continent, amplifying them, and relaying them via other satellites to different parts of the globe. U.S. Iridium satellites, for example, are launched into polar orbits and provide voice circuits, data, and paging. Their circular "spot" beams cover "cell" areas some 90 miles across, and they can communicate with each other as well as with Earth stations. They are valuable in delivering communications to and from remote areas and, along with serving the U.S. Defense Department, are used to aid heavy construction projects, emergency services, and mining industries.

GLOBAL INTERCONNECTIVITY
(Above) A geologist examines a laser positioning system installed to measure a California crater. Interpreted via GPS— a satellite-based global positioning system, which provides mapping information—measurements indicated that a spot of interest moved up 38 centimeters and northwest 21 centimeters after the Northridge earthquake near Los Angeles.

SEE ALSO

GPS

(Below) The United States' global positioning system satellites, in pink, and Russia's global navigation satellites, in yellow, continuously send signals retrievable from any place on Earth. Used to help troops in war, ships at sea, and scientists around the world, the timed digital transmissions from GPS can determine a position to within 20 yards.

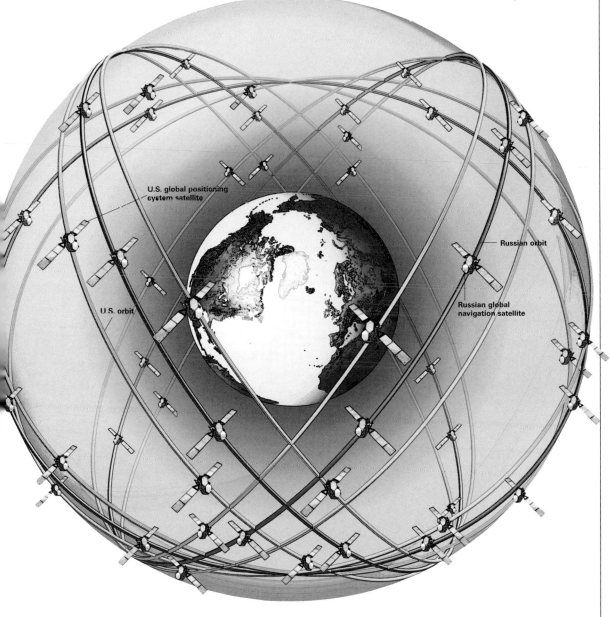

U.S. global positioning system satellite

Russian orbit

U.S. orbit

Russian global navigation satellite

PRINTING PRESSES

U NDISPUTED MASTERS OF INVENTION, THE ANCIENT CHINESE are credited with giving the world some of the earliest examples of printing. Their creations, which took the form of wood-block images on paper and silk, perhaps drew on even earlier techniques developed by the Babylonians and Sumerians, who made individual name seals.

The need for mass distribution drives modern printing. Gutenberg's invention of printing by movable type, around 1446, revolutionized the way information is disseminated, and the advent of so-called web presses and offset lithography put virtually everything fit to print into the hands of millions of readers. A newspaper's web press prints onto a continuous roll of paper as fast as 3,000 feet a minute.

Offset lithography, essential to modern printing, transfers images and text to a metal plate through photochemical action. The word "offset" means that the ink used to coat the plate does not print directly onto paper. Instead, the inked plate prints on a rubber "blanket" cylinder, and from there images are offset firmly onto paper.

Another system typesets with computer-generated text; a laser beam forms characters at the direction of the computer's electrical pulses, then transfers them to a film or light-sensitive paper. Computerized typesetting systems, which can set thousands of characters a second, also can circumvent paper by storing text and layouts as electronic files before sending them to be printed.

PRESSWORK
(Right) A six-color high-speed press blurs the action but not the quality of the print and images it leaves behind on paper.

OFFSET PRESS
(Below) Color printing using offset technology requires a complex transfer process in which paper runs over and under a meshed system of cylinders, rollers, and drums. First, a photochemical process transfers images and text to a metal plate. Rollers apply ink from troughs of cyan, yellow, magenta, and black along the line. The rollers move back and forth, spreading the ink evenly over the metal plate cylinder. The ink image then transfers onto the blanket cylinder, where paper picks up the color imprint as it passes between it and the impression cylinder below. Transfer drums move the paper to the next press, where the process repeats with another color.

SEE ALSO

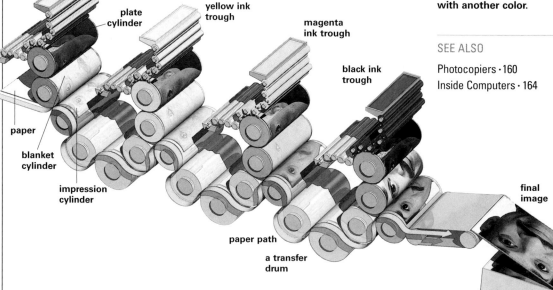

cyan ink trough

oscillating rollers

plate cylinder

yellow ink trough

magenta ink trough

black ink trough

paper

blanket cylinder

impression cylinder

final image

paper path

a transfer drum

PHOTOCOPIERS

WAXY, FINGER-SMUDGING CARBON PAPER WAS ONCE THE only way to make duplicate copies of printed material. Next mimeograph stencils, cut into coated fiber sheets by a typewriter and then wrapped around an ink-soaked drum, made copies on the paper as ink seeped through the cut outlines of the characters. Offices also had the hectograph: A master typed on special coated paper was pressed onto a large roll of gelatin-coated fabric and then peeled off, leaving its imprint in a purple, smudgy, reverse image, which printed onto sheets of paper.

In 1959, the Haloid Xerox 914 copier emerged. A relatively cumbersome apparatus, it made copies on ordinary paper. Despite problems such as paper scorching, it gave rise to the modern Xerox Corporation and transformed an industry. Today, xerography (from the Greek for "dry writing") is the standard copying technique. It works on the principle of photoconductivity: Certain materials conduct electricity under the influence of light.

A key element in today's process of photocopying is selenium, a by-product of copper refining. Selenium is a poor conductor of electricity except when it is bathed in light. Selenium elements in the copier "read" the electrical difference between a document's white and dark areas. The white areas in a document reflect light and lose electrical charge, while the dark areas do not, thus maintaining a charge. Thus the pattern on a sheet of paper is converted into an electrical version. Charged ink powder, called toner, is dusted on. Only the charged image attracts the dust, which is then fused onto the paper by heat.

MAGIC TOUCH
(Above) An office worker has only to push a button or two to produce quick and fairly inexpensive copies of a document. Centuries ago, copying was left to scribes, who carefully duplicated by hand.

SEE ALSO

INSTANT COPIES

(Below) Light from a halogen lamp illuminates a document fed into a photocopier. White areas reflect a lot of light, while print reflects little. A mirror and lens send the light onto a revolving drum with a photosensitive coating, which converts the light into a latent electrical image. Positively charged toner adheres to the negatively charged areas on the photosensitive drum; those correspond to dark areas on the original document (right). A charging electrode sends a negative charge to paper on the drum, transferring toner and the translated image onto the sheet. Finally, heated rollers fuse the image to paper.

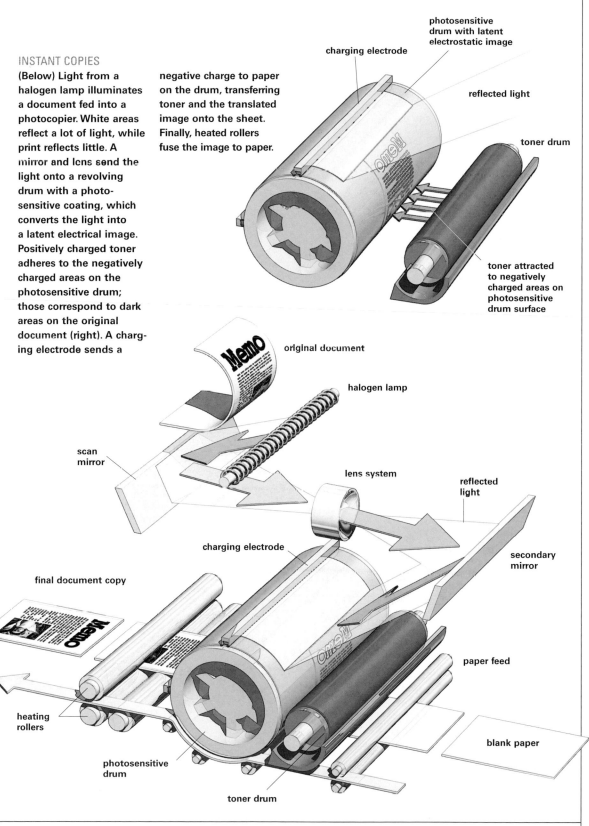

charging electrode

photosensitive drum with latent electrostatic image

reflected light

toner drum

toner attracted to negatively charged areas on photosensitive drum surface

original document

halogen lamp

scan mirror

lens system

reflected light

charging electrode

secondary mirror

final document copy

paper feed

heating rollers

photosensitive drum

blank paper

toner drum

CALCULATORS

WITH BRAINS MADE OF SILICON CHIPS CARRYING NUMBERS coded into electronic pulses, and powered by minuscule batteries or solar cells, modern calculators show astonishing speed and accuracy. They also do infinitely more than basic arithmetic. Apart from handling complicated mathematics and scientific equations, special calculators do cooking conversions, bar drink and carpentry measurements; they can crunch home financing and interest numbers and make metric and currency conversions.

A number of interrelated electronic processes are at work inside a calculator. Decimal numbers and functions are converted to a sequence of coded binary numbers that are stored on a memory chip. Instead of using the based-on-ten decimal system, the binary system counts by twos (1s and 0s only).

Next, Boolean algebra comes into play. Named after the English mathematician George Boole in the 1840s, this process involves the logic of combining numbers. Boolean principles stand behind thousands of linked microscopic switches, or logic gates, made of transistors on a chip. Labeled by Boolean nomenclature as OR, AND, and NOT, the gates evaluate all the steps of calculating by switching off for 0 and on for 1—a judgment that is the binary equivalent of false and true—and sending digital signals to appropriate locations. In units called half-adders that connect to form full-adders, the logic gates process signals so quickly that when the user presses a key, the answer appears instantaneously.

DOING THE MATH
(Below) A far cry from the pencil and paper computations once required to solve mathematical problems, calculators now do it all and in far less time. Here, a Nigerian student in Lagos presses keys to find her answer.

SEE ALSO

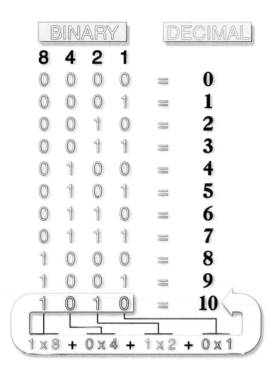

BINARY					DECIMAL
8	4	2	1		
0	0	0	0	=	**0**
0	0	0	1	=	**1**
0	0	1	0	=	**2**
0	0	1	1	=	**3**
0	1	0	0	=	**4**
0	1	0	1	=	**5**
0	1	1	0	=	**6**
0	1	1	1	=	**7**
1	0	0	0	=	**8**
1	0	0	1	=	**9**
1	0	1	0	=	**10**

$$1 \times 8 + 0 \times 4 + 1 \times 2 + 0 \times 1$$

BINARY CODE

(Left) A computer or a calculator converts information to a so-called base-2 system, a binary code that relies on 1s and 0s instead of the 10 digits of the decimal system. A number's binary code equivalent uses blocks, or positions, arranged from right to left and worth 1, 2, and powers of 2 (such as 4 and 8). Building the binary 10, for example, requires a 2-block, an 8-block, and no 4- or 1-blocks.

INSIDE A CALCULATOR

(Below) A calculator operates on the "yes" and "no," "on" and "off" principle that translates the binary code into matching voltage states. For example, when the binary system writes decimal 10 as 1010, it says the following: "Blocks of 2 and 8, yes; blocks of 1 and 4, no." In a calculator, switches see that 0s, which mean "no," or "off," are low voltage and that 1s, which mean "yes," or "on," are high-voltage. When a person presses decimal numbers on the keypad, a decoder produces the binary equivalents held in the storage cells.

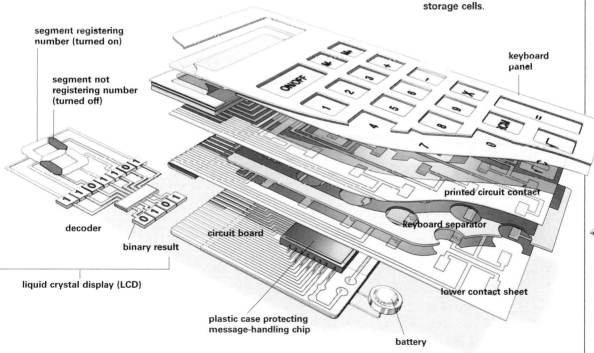

segment registering number (turned on)

segment not registering number (turned off)

keyboard panel

decoder

binary result

printed circuit contact

keyboard separator

circuit board

liquid crystal display (LCD)

plastic case protecting message-handling chip

lower contact sheet

battery

INSIDE COMPUTERS

ARLY COMPUTERS RELIED ON VACUUM TUBES—AIRLESS GLASS bottles in which electric and magnetic fields controlled the movement of electrons—to switch electrical signals and to add, multiply, store, and compare data. Developed for the radio industry, the tubes permitted machines to calculate several thousand times faster than earlier electromechanical relays. The transistor, infinitely smaller and far more frugal with power, put that speed to shame. Invented in 1947 at Bell Laboratories, the transistor is an electronic device made of semi-conducting material that carries current only under certain conditions, such as when a tiny amount of voltage is applied. Computer microchips, tiny flakes of silicon, can each contain millions of transistors linked by fine connections to form integrated circuits.

Transistors serve as electrical switches that quickly turn current off and on; this action precisely controls the flow of binary numbers, the digital data that the computer uses in its multitude of chores. Transistors govern virtually everything a computer does, interpreting and channeling digital information according to the purpose of the chip—memory, central processing, logic, or timing.

Computers use different kinds of memory within their circuitry. Random-access memory (RAM) is where program instructions and data are stored until the central processing unit—a microprocessor on a chip packed with computational controls that can move data—can access them. RAM depends on chips with memory capacity measured in accumulations of bytes, the basic units of data. (A byte is a series of eight consecutive binary digits that represents a specific alphanumeric character.)

ROM, for read-only memory (it can be read but not altered), is a permanent strongbox that stores digital information, such as start-up and operating programs, even when the computer is switched off. EPROM (erasable programmable read-only memory) may be used to store the system that the central processing unit draws on to operate start-up functions.

The more transistors there are on a chip, the faster the processing speed. Thus miniaturization has had a monumental impact on chip performance. Transistor designers are also researching so-called quantum mechanical versions, which have been likened to traveling through a tunnel in a mountain that would be impossible to climb. The devices essentially enable electrons to "tunnel" under energy barriers to get to places that would appear impossible to reach. Such a shortcut could mean an increase in speed, far lower power needs than in conventional transistors, and effective operation with far fewer transistors than are now required in an integrated circuit.

CIRCUIT BOARD
(Right) Repository for computer chips and other interconnected electronic components, a circuit board is a model of intricacy. Tiny copper pathways are impressed on a fiberglass or plastic base. Along those pathways speed electrons, the basic component of information manipulated by every computer. Computers have one or more boards that are known as cards, chief among them the motherboard, or system board, which contains the central processing unit (CPU), memory, and other operating elements.

SEE ALSO

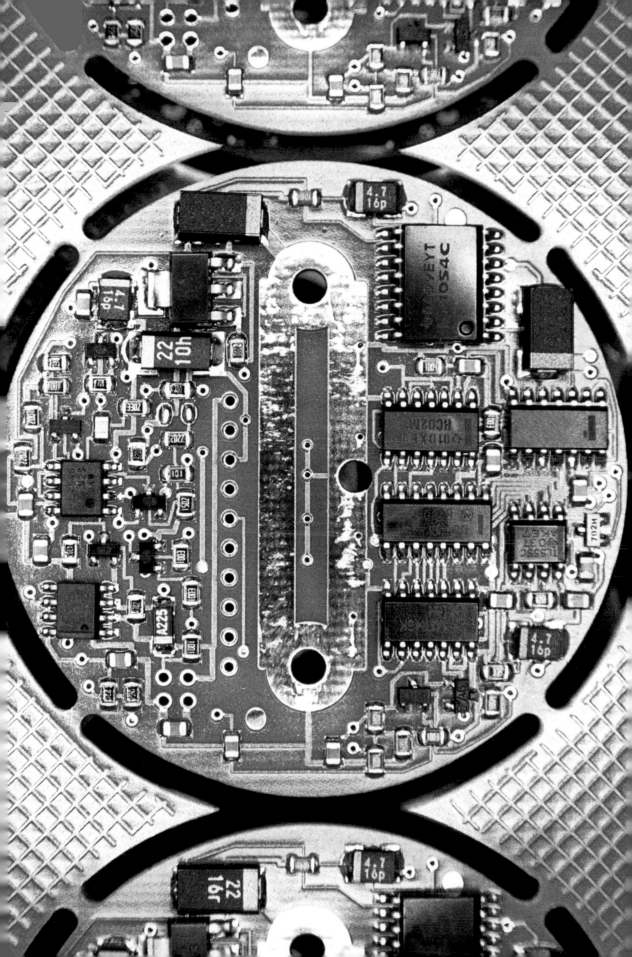

THE INTERNET

WHEN THE INTERNET WAS CREATED IN 1969 AS A U.S. DEFENSE Department program known as ARPANET—Advanced Research Projects Agency Network—its goal was to establish a secure communications system that would connect computers between different locations and survive in the event of hostilities. Soon researchers and academics began to use the system. Today the Internet is many networks arranged like a gigantic spider's web throughout the world. No one corporation or governmental agency owns it: It is publicly accessible, with many millions of people plugged into its World Wide Web, labeled "www." While Internet communications traffic still travels over standard telephone lines to some extent, it also moves through cable-television lines and via satellites, fiber optic links, and radio signals.

At the heart of the system are the Internet Service Providers (ISPs), computerized links to the Internet. Chains of ISPs communicate through public network access points (NAPs, which cover a broad geographic region) and local access points, called LAPs. A phone line or one of the other transmission modes grants access to an ISP, while a browser—a software program that connects with a server—allows the user to view and navigate, or surf, the Internet. To help locate information there are search engines, which connect to enormous databases.

Data available on the Internet are broken down electronically into "packets," which may contain coded address information, and are sent out from server to server, moving through exchange or access points. But in order to accommodate a request for information, servers must send the electronic information to the right "computer client"; moreover, once they get there all the data contained within have to be collected and reassembled. All of this is the job of the communications protocols, which are rules followed by the servers. The most important of these are the TCP, the Transmission Control Protocol, which collects and reassembles the fragments of data, and the IP, the Internet Protocol, which handles routing. Protocols employ various forms of cyberspeak to get their work done, like the familiar http://, for "hypertext transfer protocol," which is one of the most important "languages" used in addresses and to move files on the World Wide Web. Other protocols translate electronic mail (e-mail) and direct file transfers, identify the origin of information, and allow one to log onto, or determine who's logged onto, a remote computer.

Web addresses have other essential components: a specific domain name and the "dot" parts that follow it. When complete, a Web address is called a URL (Universal Resource Locator). Typing in a combination of, say, the protocol http:// with a unique domain name on the computer alerts the servers of the request and where to send it.

ALWAYS IN TOUCH

(Right) Seated in a tent atop Tahir Tower in Karakoram, Pakistan, a woman reads and delivers her e-mail as casually and efficiently as she would in an office. Easily accessible with simple telephone jacks and cord, or using wi-fi (wireless fidelity) Internet access, the World Wide Web's more than 92 million gigabytes of data (enough to fill a two-billion-volume encyclopedia) inform, educate, and entertain countless users. Perhaps the most valuable tools of the information age, the Web and the Internet it draws on have changed the way the world obtains, compares, and keeps track of knowledge and news.

SEE ALSO

NETWORKING

(Below) A typical network connects businesses, homes, universities, and government agencies, allowing access to information. Powerful gateway computers operated by service providers connect different wide-area networks to one another, providing Internet access and a host of services, relaying digitized information, and making each network's computers able to communicate with one another. Satellites and phone lines relay data and messages between computers, each one equipped with a modem or other device for transmitting and receiving data. Routing computers along the way decode instructions on the transmissions that tell them how and where to send the messages. Conceived in 1969 by the Advanced Research Projects Agency at the U.S. Defense Department, the first nationwide network linked computers at four universities.

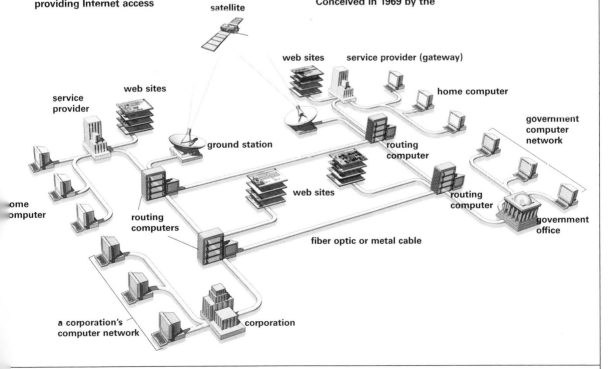

BAR CODES & SCANNERS

USING COMPUTER TECHNOLOGY'S OMNIPRESENT SERVANT, the binary system, bar codes represent the decimal price and other details about an item in a series of parallel vertical lines and white spaces. Such codes also are used to track documents, packages, gene sequences in databases, and books in libraries and bookstores. In general, bar codes are standardized. Bar codes in stores come under the Uniform Product Code, which assigns a unique numerical signature to grocery products; bookstores use codes set by the International Standard Book Number (ISBN) system, which identify a book's price and the country that published it.

When a cashier passes the item over (or under) a scanner, a laser beam set to a specific frequency reads the binary code in the bars; the information is transmitted to a decoder and a computer, and the computer searches the files and locates the item and its price. The information may then be sent to the store's main computer as sales and inventory information, and at the same time it informs the computer in the cash register, which adds it to the total of items purchased and displays its price on the screen.

BAR CODES

(Right) No cash register bells ring as a scanner tracks a package via its bar code. Introduced in the 1970s, bar codes identify products and prices. Businesses that must track inventory rely on barcoding; so do makers of identification cards and researchers who monitor gene banks. Scanners that read the codes come in various models, handheld or built into counters.

SCANNING

(Left) When a clerk draws a bar-coded item over a scanner in a checkout counter, a laser beam enters a spreader and reflects off a mirror to a disc and up through the opening. The beam reads the encoded information and bounces back to the scanner's laser detector; the signal then goes to a computer that displays the price. The bars of the code represent the item in binary digits 0 (in white), which reflect more strongly, and 1 (in black).

SEE ALSO

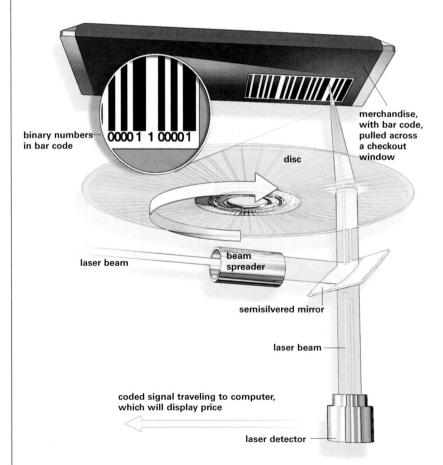

binary numbers in bar code

0000 1 1 00001

merchandise, with bar code, pulled across a checkout window

disc

laser beam

beam spreader

semisilvered mirror

laser beam

coded signal traveling to computer, which will display price

laser detector

THE STOCK EXCHANGE

A STOCK EXCHANGE IS AN OFTEN CHAOTIC PLACE—AN AGENCY auction market where the securities of corporations and municipalities, such as stocks and bonds, are bought and sold daily. The world's largest exchange is New York's, with roots going back to 1792 on Manhattan's Wall Street. Some 3,000 companies are listed on the so-called big board, and hundreds of brokers, support staff, and various specialists man the trading floor, handling a daily volume of one billion shares.

Once, the only technology that graced a trading floor or the offices of brokers and corporate executives was the stock ticker, a glass-enclosed wooden pedestal through which stock quotes chattered, printed on streams of paper tape transmitted by the Western Union Telegraph Company. Today, electronics so rule the New York Stock Exchange that what was a whole day of trading ten years ago is now handled in the opening half hour, sometimes even in the first 15 minutes; a central computer can fill orders at the astonishing rate of more than 200 transactions a second.

The typical trading order goes through a series of computer-assisted steps from the time a customer places it with a broker by phone or through computer access, to when it reaches the floor of the Exchange. Some customers skip a broker and trade directly on the Internet. On the floor, an electronic order-routing system stores the order, then sends it to a screen at a broker's booth—or directly to a trading post, where specialists for particular stocks work—for speedy electronic processing and execution of the deal.

Computers also track the value of groups of stocks, and allow the broker to compare stock prices with those on other exchanges and compete for the best buys; automation also sends changing stock values to the familiar scrolling "ticker" that crawls across the bottom of a television screen or on the big boards of brokerage offices.

A wireless data system enables brokers to get orders and send reports with mobile handheld devices from anywhere on the Stock Exchange floor, and 3-D data displays can be accessed on the Web to bring the Exchange's floor action to investors. It is easy to understand why 99 percent of all orders are delivered to the point of sale electronically.

TRADING FLOOR

(Right) A mecca for brokers and investors, the New York Stock Exchange is the world's largest and most technologically advanced equities market. The Exchange uses advanced computer technologies to help shrewd individuals and investment novices alike make decisions based on minute-by-minute information.

SEE ALSO

Sending Signals · 118
Cell Phones · 154
Inside Computers · 164
The Internet · 166

KEEPING STOCK

Illuminated stacks of numbers keep the focus of brokers on the electronic display at stock exchanges. Such numbers light up around the world, not only in stock exchanges but also on television, computer screens, and in displays in local brokerage offices. The boards reflect market fluctuations, or changes in the prices of various stocks and bonds. Stock markets differ by country in the way they connect with their governments. The New York Stock Exchange, for instance, is not operated by the government but is regulated by United States law, while the Shanghai Stock Exchange is governed by the China Securities Regulatory Commission.

CHAPTER 10

OTHER WORLDS

TOOLS THAT LET US OBSERVE REALMS TOO SMALL OR too distant to be seen by the naked eye are often baffling. Even the familiar optical devices—contact lenses, bifocals, and binoculars, for example—work in ways we may only vaguely understand. Employing lenses, mirrors, and prisms, the precision instruments we use to view microscopic or macroscopic worlds may have direct or indirect links to laws of reflection and refraction (which deal with changes in the direction of light as it passes from one medium to another). Some devices follow different laws, using lenses made from electromagnets to focus or change magnification and beaming electrons rather than light to study objects. Others have no lenses at all and look nothing like conventional telescopes. Called radio telescopes, these dish-shaped devices receive radio waves from space, allowing us to observe other worlds, near and far, in yet another way.

Made transparent as its opening turned during the camera exposure, the Keck Telescope dome graces the top of Hawaii's Mauna Kea.

LENSES

PTICS IS THE SCIENCE OF LIGHT AND VISION. OPTICAL LENSES are devices that aid vision by focusing, bending, and spreading light rays emitted or reflected by an object. The lens in a human eye can shift focal length through muscular contraction and relaxation. Optical lenses have focal length and power built into their shape. They generally consist of two curved surfaces, or one flat and one curved; these curves may be concave (inward curving) or convex (outward curving). Surfaces and thicknesses determine a lens's focal power and function; combinations of different lenses—the compound lenses cemented together in a basic light microscope, for example—prevent the blurring, distortion, and other anomalies that can occur with single thin lenses.

Eye specialists have many techniques for measuring the eye's refractive power and for determining a person's need for corrective lenses in eyeglasses or in implants like those used after cataract surgery. To compensate for nearsightedness, lenses are ground in concave shapes; for farsightedness, lenses are convex. Cylindrical lenses are used for astigmatism, a condition in which light does not focus properly on the retina because of a defect in the curvature of the natural lens.

In bifocals, the upper and lower parts of a lens are ground differently, to correct for both close and distant vision. Trifocals are ground with a center lens for intermediate distance. Intraocular cataract lenses that replace clouded natural lenses are made of a variety of synthetic elastic materials. They may be monofocal lenses, which provide vision at a fixed distance, or multifocal, which broaden viewing distance near to far.

MAGNIFYING GLASS
(Right) A piece of high-tech equipment receives a visual inspection in a modem manufacturer's quality-control lab. Complex creations in convex and concave forms, lenses compensate for inadequacies in human eyes.

BINOCULAR VISION
(Below) To see, human eyes form two images of an object, one on each retina. Binoculars accommodate that by letting light enter two large objective lenses. Reflecting prisms inside each tube fold the light's path, providing higher magnification than that afforded by field glasses, which have no prisms. The prisms also allow for a more compact binocular design; field glasses need much longer and heavier tubes for high magnification.

SEE ALSO

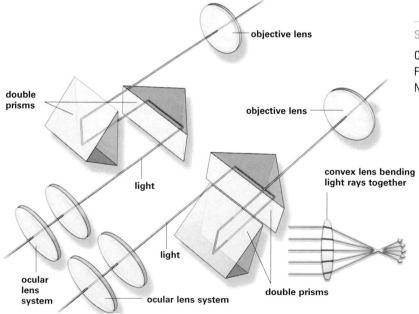

objective lens

double prisms

objective lens

light

convex lens bending light rays together

ocular lens system

light

ocular lens system

double prisms

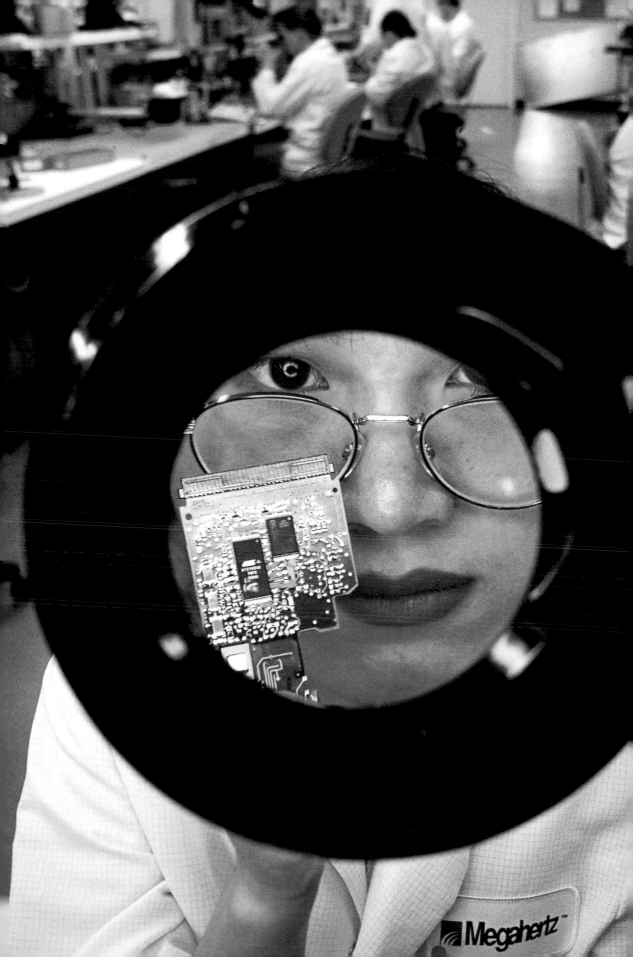

NIGHT VISION

W HAT DOES IT TAKE TO SEE IN THE DARK? HUMANS AND other mammals need the rods in the retina, millions of tiny cylindrical elements that contain rhodopsin, a purple pigment that can detect dim light. We also have millions of retinal cones, light-sensitive cells that enable us to read fine print. Cats and owls have only rods, which makes them nocturnal creatures capable of seeing far better than humans can in low light.

For a human eye to see in a dark movie theater, its rods have to rejuvenate rhodopsin, which is so chemically altered by bright outside light that the rods temporarily lose their sensitivity. While rhodopsin's response allows the eye to withstand the sunlight, it does not help much when you enter a dark screening room. Gradually, however, more rhodopsin is produced, increasing sensitivity to the theater's low light. Vitamin A, a lack of which can result in night blindness, is essential to the production of rhodopsin.

Medical science may someday create a pill that allows us to see better in the dark, but until that time, technology must fill the bill. Night-vision binoculars and scopes are electro-optical instruments that are incredibly sensitive to a broad range of light, from visible through infrared. Light that enters a lens in a night-vision scope reflects off an image intensifier to a photocathode and then is converted to an electronic image. Amplified on a viewing screen, the image reveals much more than a night scene observed through a conventional scope.

NIGHT TRAINING
(Right) Night-vision technology enables Israeli soldiers to peer through darkness.

ELECTRO-OPTICS
(Below) To intensify available light, a photocathode accelerates photons through a vacuum. Then a microchannel plate focuses the image on a phosphor screen.

NIGHTSIGHT
Certain night-vision devices, including night goggles, require some ambient light, even if it comes from the stars or from a focused-beam infrared source. Like cameras, night-vision devices offer the user various image magnification choices, which is an especially valuable feature during nighttime military operations involving close air support of ground troops. Technology within the system converts photons into electronic images and focuses them on a phosphor screen. The images are enhanced with amplified green light, creating an eerie but visible picture of a soldier's surroundings.

SEE ALSO

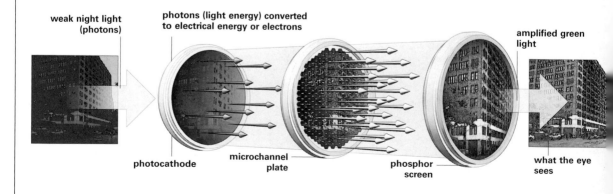

weak night light (photons)

photons (light energy) converted to electrical energy or electrons

amplified green light

photocathode

microchannel plate

phosphor screen

what the eye sees

MICROSCOPES

"WHERE THE TELESCOPE ENDS," WROTE VICTOR HUGO IN THE novel *Les Misérables*, "the microscope begins. Which of the two has the grander view?"

As nearly everyone knows, microscopes are instruments that give us a glimpse into the world too small to see with the naked eye. But when Victor Hugo penned those words in the 19th century, he was speaking of the simple light microscope with its then astonishing—and now very limited—view. Today, the telescope has not ended at all, and the microscope has long said good-bye to its beginning.

Perhaps most familiar is the ordinary compound microscope, which has a pair of convex lenses of short focal length on the lower end of a tube, called the objective, and another pair at the eyepiece. When a specimen is placed at the objective end and is illuminated with light reflected from a mirror, the magnified image is magnified again at the eyepiece end. Focusing is done by moving the objective lenses nearer to, or farther from, the specimen.

While the light microscope can identify the form and structure of extremely tiny organisms, the electron microscope reveals far more detailed information about their surface and inner workings. Under an electron microscope, for example, a simple bacterium bares its very soul and becomes an intricate cutaway revealing all the complexities of life.

With a magnification power of up to hundreds of thousands of times the size of the original, the electron microscope can see objects that are among the most invisible of the invisible: atoms. Armed with an electron gun, this nonlight microscope focuses a beam of electrons through a vacuum and over the surface of a specimen. A signal is generated, projected onto a fluorescent screen, and then photographed.

The electron microscope uses magnetic lenses to focus and change its level of magnification. These produce a field that acts on the electron beam in the same way a glass lens works on light rays. The exceptionally high resolution is due to the shorter wavelength associated with electron waves.

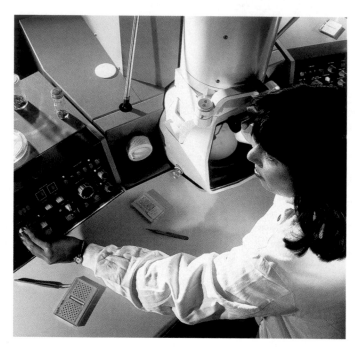

TV TECHNOLOGY

(Below) An electron microscopist gets finer details than through a light microscope. Using electron beams, not light, and a monitor, not an eyepiece, for viewing, the electron microscope draws more on television technology than on lens science for its principles of operation. Electromagnets, not glass, function as the lens. Electrons pass through thinly sliced specimens coated with gold or water vapor to improve the image.

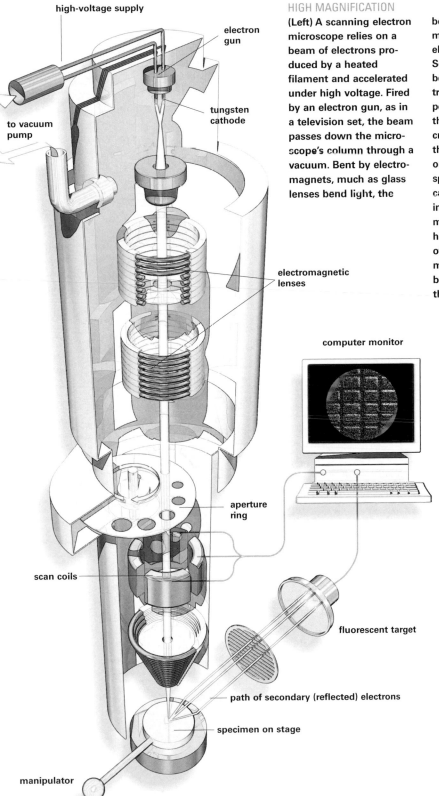

high-voltage supply

electron gun

tungsten cathode

to vacuum pump

electromagnetic lenses

computer monitor

aperture ring

scan coils

fluorescent target

path of secondary (reflected) electrons

specimen on stage

manipulator

HIGH MAGNIFICATION

(Left) A scanning electron microscope relies on a beam of electrons produced by a heated filament and accelerated under high voltage. Fired by an electron gun, as in a television set, the beam passes down the microscope's column through a vacuum. Bent by electromagnets, much as glass lenses bend light, the beam focuses on a specimen cut thin to allow the electrons to penetrate. Scanning coils sweep the beam in a gridlike pattern, training it on specific points. As the beam passes through the specimen, it creates electrical signals that appear as images on a monitor. For thick specimens, microscopists can use higher accelerating voltages. High magnification—by hundreds of thousands of times—and different models of telescope may be used, depending on the specimen.

TELESCOPES

GALILEO DID NOT INVENT THE TELESCOPE, BUT HE WAS PROBABLY the first person to use one for making serious astronomical observations. With his 17th-century device, he found moons around Jupiter, mountains on Earth's moon, and spots on the sun. Galileo's telescope was a simple arrangement of lenses in a tube that bent light at odd angles, thus distorting color and producing images of poor quality. Later, Isaac Newton rectified that, using mirrors to reflect light rather than bend it. Modern astronomers often view an image on a screen or as a photograph, without actually looking through an eyepiece.

In telescopes used in observatories, gigantic mirrors of polished glass can have diameters of up to 32.2 feet. Spun-cast in rotating furnaces from tons of raw glass, they are linked to computers and photodetectors that, in turn, link multiple telescopes for greater viewing power. The mirrors have enormous light-gathering power and can view wide areas of the sky for large-scale surveys of the faintest objects in deep space. Indeed, instruments now under development will allow simultaneous observation of the spectra from as many as 300 galaxies. Just as important, cast from vastly improved glass mixtures and finely polished by beams of electrically charged atoms, they are distortion-free.

Computers also eliminate atmospheric distortion by analyzing light and correcting it if necessary. New optical detectors promise even more: Their sensitivity allows them to clock the arrival of a single light particle and to measure its energy with exceptional precision—all through the infrared, optical, and ultraviolet segments of the spectrum.

SKY DOME

(Right) Inside the dome of each twin Keck Telescope atop Hawaii's 13,600-foot-high Mauna Kea, a 32.2-foot-in-diameter mirror contains 36 hexagonal segments, which can be aimed by computer. The world's largest optical and infrared telescopes, the Keck telescopes stand eight stories tall, and each weighs 300 tons.

REFLECTING TELESCOPES

(Below) In these telescope types, light passes into the main mirror and focuses the image on a secondary mirror. Astronomers today tend to use either Cassegrain or Coudé focus designs.

SEE ALSO

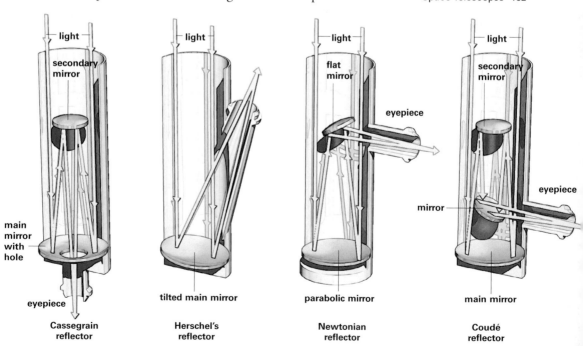

Cassegrain reflector — light, secondary mirror, main mirror with hole, eyepiece

Herschel's reflector — light, tilted main mirror

Newtonian reflector — light, flat mirror, eyepiece, parabolic mirror

Coudé reflector — light, secondary mirror, mirror, eyepiece, main mirror

SPACE TELESCOPES

WHILE GROUND-BASED OBSERVATORIES KEEP EXPANDING their ability to see into deep space, there is, as the saying goes, nothing like being there. Orbiting observatories, for example, can see into radiation wavelengths difficult to image by Earth-based instruments, or examine high-energy processes in the nuclei of galaxies and in the vicinity of black holes.

One of the most powerful space telescopes bears the name of Edwin Powell Hubble, the U.S. astronomer who discovered that certain nebulae are really galaxies outside our own Milky Way. The Hubble Space Telescope, launched aboard the space shuttle *Discovery* in 1990, has sent back images that prove, for example, that the entire universe is

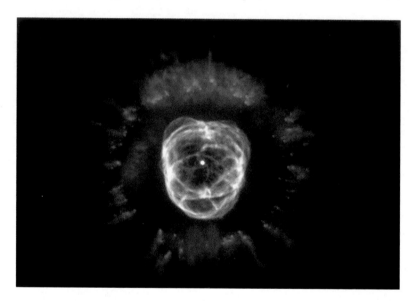

expanding, rushing outward from what may have been the event commonly called the big bang of creation.

The Hubble was designed to gather light from a large, concave primary mirror and reflect it off another mirror into an array of sensors. Sensitive instruments detect x-rays, infrared light, and ultraviolet light, revealing the makeup of far-off celestial structures and systems. By focusing on a seemingly empty bit of sky, Hubble even discovered what might be called a construction site for galaxies: a vast birthplace of stars, 12 billion years in the past, that had been invisible to astronomers until the Hubble could travel closer and amplify the view.

Now a replacement, the new James Webb Space Telescope, named after NASA's second administrator, is scheduled for launch in 2011 for a three-month journey into orbit. It will have a larger mirror, to give it more light-gathering power; it will operate much farther from Earth— a million miles away; and it will see deeper into space than the Hubble.

THE BIGGER PICTURE

(Right) With its huge mirrors, cameras, and spectrographs seeking, and sometimes finding, distant glimmers from the big bang, the Hubble Space Telescope converts faint starlight into visible limelight. It orbits 375 miles above Earth, providing images free of the distortion seen in views from Earth-based telescopes not equipped with computer-controlled adaptive optics. The telescope's imaging spectrographs can read light emitted by gases, depicting it as red for nitrogen, green for hydrogen, and blue for oxygen.

STELLAR RELIC

(Left) Captured by the Hubble Space Telescope and dubbed the Eskimo Nebula—because to some it resembles a face framed by a fur parka—this glowing disk is what's left of a dying sunlike star, 5,000 light years away from Earth in the constellation Gemini. The "parka" is actually material embellished with a ring of comet-shaped objects.

SEE ALSO

Fiber Optics & DSL · 152
Telescopes · 180

ILLUSTRATION CREDITS

COVER: Royalty-Free/CORBIS.

INTRODUCTION: 2-3, Steven Gottlieb/Getty Images; 7, Donald Brewster/Bruce Coleman, Inc.; 8, Dan McCoy/Rainbow.

CHAPTER 1. AT HOME: 12-13, Andrew Michael/Getty Images; 15, David Arky/CORBIS; 16, Rita Maas/Getty Images; 19, A. Syred/Photo Researchers, Inc.; 20, Tom Stewart/CORBIS; 22, Wally McNamee/CORBIS; 25, Susan Lapides; 26, Michael Freeman; 29, Matthias Kulka/zefa/CORBIS.

CHAPTER 2. POWER & ENERGY: 30-31, Jim Zuckerman/CORBIS; 33, Royalty-Free/CORBIS; 35, Peter Essick; 37, Roger Ressmeyer/CORBIS; 38, Derek Croucher/CORBIS; 41, Roger Ressmeyer/CORBIS.

CHAPTER 3. BUILDING: 42-43, Erik Leigh Simmons/Getty Images; 45, Joel Rogers/Getty Images; 46, Janet Gill/Getty Images; 49, Robert Essel NYC/CORBIS; 51, QA Photos; 53, Roger Ressmeyer/CORBIS.

CHAPTER 4. TRANSPORTATION: 54-55, Sam Abell; 56, John Madere/CORBIS; 59, Ruaridh Stewart/ZUMA/CORBIS; 60, Chris Trotman/Duomo/CORBIS; 62, Dan Lamont/CORBIS; 64, Richard A. Brooks/AFP/Getty Images; 65, Charles O'Rear; 66, Stuart Westmorland/CORBIS; 69, Onne van der Wal/CORBIS; 71, Stuart Westmorland/CORBIS; 72, Tom Tracy/CORBIS; 74, Bruce Burkhardt/CORBIS; 76, Bettmann/CORBIS; 78, Erik Viktor/Science Photo Library/Photo Researchers, Inc.; 79, NASA.

CHAPTER 5. MATERIALS: 80-81, Spencer Jones/GlasshouseImages.com; 83, Mark Douet/Getty Images; 85, Davies & Starr, Inc.; 87, David Stoecklein/CORBIS; 88, Phil Schermeister/NGS Image Collection; 89, James L. Amos/CORBIS; 91, Richard Cummins/CORBIS; 93, Roger Tully/Getty Images; 95, Antonio Mo/Getty Images.

CHAPTER 6. MANUFACTURING: 96-97, Bryan F. Peterson/CORBIS; 99, Peter Dean/Getty Images; 101, Pierre Vauthey/CORBIS SYGMA; 102, Joel Sartore; 104, Peter Essick; 106, Digital Vision/Getty Images; 108, John Lawrence/Getty Images; 111, Charles O'Rear/CORBIS; 113, Kelly-Mooney Photography/CORBIS.

CHAPTER 7. ENTERTAINMENT: 114-115, Evan Agostini/Getty Images; 117, Randy Duchaine; 118, Roger Ressmeyer/CORBIS; 119, Cristi Matel/Shutterstock; 121, Penny Tweedie/CORBIS; 122, Dr. Jeremy Burgess/Photo Researchers, Inc.; 124, Tim Boyle/Getty Images; 127, Steve Prezant/CORBIS; 128, Tim Laman; 130, Lucasfilm Ltd./20th Century Fox, courtesy Photofest; 133, Robert Landau/CORBIS; 135 Tony Wiles/Getty Images.

CHAPTER 8. HEALTH & MEDICINE: 136-137, Dr. Arthur Tucker/Photo Researchers, Inc.; 138, Allan H. Shoemaker/Getty Images; 141, Collection CNRI/Phototake; 143, Scott Camazine/Photo Researchers, Inc.; 145, Lester Lefkowitz/CORBIS; 147, Tom Stewart/CORBIS; 148, John Zich/Time Life Pictures/Getty Images; 149, Joe McNally.

CHAPTER 9. INFORMATION: 150-151, Charles O'Rear/CORBIS; 153, Lawrence Manning/CORBIS; 155, Edward Bock/CORBIS; 156, Roger Ressmeyer/CORBIS; 159, B.C. Moller/Getty Images; 160, Michael Malyszko/Getty Images; 162, James Marshall/CORBIS; 165, Samuel Ashfield/Getty Images; 167, Jimmy Chin/NGS Image Collection; 169, Jon Feingersh/CORBIS; 171, Gail Mooney/CORBIS.

CHAPTER 10. OTHER WORLDS: 172-173, Roger Ressmeyer/CORBIS; 175, Joel Sartore; 177, Shaul Schwarz/CORBIS; 178, Geoff Tompkinson/Science Photo Library/Photo Researchers, Inc.; 181, Roger Ressmeyer/CORBIS; 182, NASA; 183, NASA.

ACKNOWLEDGMENTS

The Book Division wishes to thank the many individuals, groups, and organizations mentioned or quoted in this publication for their help and guidance. The current edition is the third revision of a book originally published in 1983.

For their help on the first *How Things Work,* the editors express gratitude to Andrew J. Pogan, the book's chief scientific consultant; Keith Bryan; Neil W. Averitt; Edmond P. Bottegal, Joy Mining Machinery; Dennis R. Dimick; Clarence E. Hill; Brenda Hooper of Arc Second; Sandy Miller Hays, USDA ARS Information; Mark Princevalle, Nuclear Communications Services, Northeast Utilities System; George Sanborn, Massachusetts Transportation Museum, Boston; and Christopher Stewart.

For the present edition, we greatly appreciate the keen eye and thorough knowledge of Hal Hellman, who served as our expert reader and consultant in science and technology. He is the author of *Great Feuds in Mathematics, Great Feuds in Science,* and 28 other books on science, technology, and related topics. He has published articles in the *New York Times, Omni, Reader's Digest, Psychology Today,* and *Geo;* has taught science writing at New York University; and has lectured across the country.

Author John Langone, a veteran science journalist, was a staff writer for *Discover* and *Time* magazines, a reporter and writer for United Press International, and science editor at the *Boston Herald.* He was a Kennedy Fellow in Medical Ethics at Harvard, a Fellow at the Center for Advanced Study in the Behavioral Sciences at Stanford, and a Fulbright Fellow at the University of Tokyo, where he researched the impact of science and technology on the Japanese. He wrote a weekly column, "Books on Health," in the science section of the *New York Times.* This was Langone's 25th book.

Sadly, John Langone died in 2006, while this edition of his book was nearing production. We trust that in our revision, we have respected him in word and spirit.

GLOSSARY

Airfoil: The curved surface of a wing or propeller that produces lift as it moves through the air.

Alloy: A substance composed of two or more elements, usually two or more metals or a metal and a nonmetal melted together.

Amplitude: The strength and size of a radio wave.

Anneal: To heat glass, metal, or an alloy and then let it cool gradually to room temperature, thus making it less brittle.

Aquaculture: The breeding and raising of fish and seafood under precise conditions to improve food production and increase the ability to stock streams, ponds, and lakes.

Aqueduct: An artificial channel or structure that conveys water.

Arch: A curved construction element that spans an opening and supports the weight above it.

Atomic clock: A precision clock that operates by measuring the natural vibration frequencies of atoms or molecules.

Bar code: Coded information that identifies a labeled object. The bars and spaces (and sometimes numerals) are designed to be scanned and read by a computer.

Bessemer process: A steelmaking process that removes impurities from pig iron by forcing a blast of air through molten metal.

Binary code: A base-two number system that uses only 1s and 0s, unlike the standard base-ten, or decimal, system.

Bobbin: A spool on which weft yarn is wound for weaving, or on which thread is wound for use in a sewing machine.

Buoyancy: The lifting force of a fluid exerted on a body immersed in it.

Buttress: A projecting masonry structure that supports or strengthens a wall or building.

Caisson: A watertight chamber used in the construction of foundations, bridges, and tunnels.

Cam: A sliding or rotating device used to change rotary motion to linear motion, or vice versa.

Cast iron: A hard and brittle iron alloy consisting of carbon and silicon cast in a mold.

CD-ROM (compact disc read-only memory): A device for storing computer data in digital form.

Celluloid: A tough, flammable thermoplastic composed of nitrocellulose and camphor.

Centrifugal force: The force that presses objects away from the center of rotation.

Chlorofluorocarbon: A compound containing carbon, chlorine, fluorine, and sometimes hydrogen. It is used in refrigerants, cleaning solvents, aerosol propellants, and plastic foams.

Combine: A multifunction farming machine that cuts, threshes, cleans, and gathers cereal crops.

Compression: A force that reduces or shortens a material by pressing or squeezing it together; the opposite of tension.

Compressor: The part of a cooling system in which the coolant, in its gaseous state, is compressed by a pump to heat it.

Concrete: Strong building material made by combining sand, gravel, rock, a cementing material, and water.

Condenser: In a cooling system, the part in which the coolant releases heat and changes from a gas or vapor into a liquid.

Cone: A photosensitive receptor cell in the eye's retina.

Contour plowing: Plowing across the fall of the slope rather than with it. This process helps control water runoff and soil erosion.

CPU (central processing unit): The primary chip in a computer. It not only controls the operation of other components in the computer but also interprets and executes instructions.

Curtain wall: A building's nonbearing exterior wall.

Defibrillator: An electronic device that uses electric shock to restore rhythm to a rapid and irregular heartbeat.

Density: The weight of an object or a substance in relation to its volume.

Digital: Relating to data in the form of numerical digits.

Diode: An electronic component that lets current flow through it in only one direction.

Drag: The force with which air or water resists the motion of a moving body.

Electricity: A natural entity composed of electrons and protons (or possibly electrons and positrons). It is observable in bodies electrified by friction, in lightning, and in the aurora borealis; it is generally used in the form of electric current.

Electromagnet: A magnet using electricity to increase its power.

Electromagnetic radiation: The form in which energy moves through space or matter.

Electromagnetic spectrum: The full range of light energy from the sun, consisting of electromagnetic waves distinguished by their frequency, or wavelength.

Electron: A subatomic particle with a negative charge.

Endoscope: A flexible instrument for seeing inside organs of the human body.

Energy: The capacity to do work.

Evaporator: The part of a cooling system where coolant absorbs heat and changes from liquid to gas or vapor.

Filament: A metallic wire in an incandescent lamp that glows when an electric current passes through it.

Fission: The splitting of an atom's nucleus to release large amounts of energy.

Flying buttress: A projecting arched structure that carries the lateral thrusts of a roof or vault to an upright pier or buttress.

Focal length: The distance from the surface of the lens to the focal point.

Force: Any influence that exerts pressure on an object or that speeds up, slows down, or stops an object's motion.

Fossil fuel: A fuel formed from the fossilized remains of plants and animals.

Frequency: The number of wave vibrations over a given period of time.

Fuel cell: A device that changes the chemical energy of a fuel and an oxidant directly into electrical energy.

Fusion: A nuclear reaction in which the nuclei of light atoms, such as hydrogen, come together and form one or more heavier nuclei, thereby releasing enormous energy.

Gamma ray: A form of electromagnetism that has the shortest known wavelength. It is also the most penetrating form of radioactivity.

Gear: A mechanism using toothed wheels to convey movement from one part of a machine to another.

Generator: A device that converts mechanical to electrical energy.

Geothermal energy: Energy obtained from subsurface reservoirs of naturally occurring heat.

Gnomon: A sundial's pointer, which indicates the time of day by the position or length of its shadow.

Hard disk: In a computer, a data-storage device that uses a magnetically coated metal disk.

Harness: A frame that holds and controls heddles, the parts of a loom that hold the warp threads.

Herbicides: Chemical preparations that are used to destroy vegetation, usually weeds.

Hertz: One hundred cycles of electromagnetic radiation per second.

Horsepower: A unit of power equal to 745.7 watts; in metric (CV), the power required to raise 75 kg one meter in one second, or 735.5 watts.

Hybrid: The product of two animals or plants having different gene compositions.

Hydrocarbons: Organic compounds containing only carbon and hydrogen.

Hydroponics: A type of soilless agriculture in which plants are grown in solution or in a moist inert medium that provides the necessary nutrients.

Hypersonic: Having a speed of at least five times that of sound.

Infrared ray: A form of electromagnetic energy whose wavelength is slightly larger than that detected by human eyes.

Internet: A worldwide system of computer networks linked by communication systems.

Isotope: An atom of a chemical element having a different number of neutrons, giving it a different mass number.

Kilowatt-hour: A commercial unit—the amount of work or energy equal to one kilowatt expended in one hour.

Kinetic energy: The energy generated by an object's motion.

Laparoscope: A flexible fiber-optic instrument inserted through an incision to visually examine the abdomen.

Lesion: An abnormal change in a part of the body caused by disease or injury.

Lift: The upward force produced by an aircraft's wing, a helicopter's rotor, or a hydrofoil's foil as it moves.

Lintel: A horizontal structural member across the top of an opening. It carries the weight of the wall above it.

Live load: The total weight of a structure, including its contents and occupants.

Load: The force on a structure caused by weight or wind pressure.

Maglev (magnetic levitation): A technology that uses the physical properties of magnetic fields generated by superconductors to make a vehicle float above a solid surface.

Megahertz: One million cycles of electromagnetic radiation per second.

Microprocessor: A single computer chip that has a complete central processing unit.

Microwaves: Electromagnetic radiation used in radar and microwave ovens.

Monomer: Small molecule that is the basic building block of a polymer.

Optical fiber: Fine strand of glass capable of transmitting digital information in the form of light pulses.

Pesticide: A chemical used for killing weeds, insects, and fungal diseases.

Petrochemical: A chemical that is derived from petroleum or natural gas.

Photocathode: A device that converts light waves to electrical charges.

Photocell/photoelectric cell: An electronic device that

modifies the flow of an electric current in response to light waves.

Photodiode: A semiconducting device that converts light pulses to electronic signals.

Pier: A supporting column, pillar, or pilaster designed to take vertical loads.

Piezoelectric reaction/piezoelectricity: The electric current or electric polarity produced by applying pressure to a crystalline substance, such as quartz.

Pig iron: Iron produced in a blast furnace and used to make steel, cast iron, or wrought iron.

Pile: A long, slender column of steel, concrete, or wood driven into the ground to support the vertical load of a sructure.

Pinion: A gear having a small number of teeth that mesh with teeth on a larger gear.

Pitch: In aeronautics, the angle at which a body slants on the axis running from front to back; in music, the tone perceived coming from a specific vibrational frequency.

Pivot: A shaft or pin on which a lever turns.

Pixel: The acronym for "picture element," the smallest unit, or dot, of color and brightness in an electronic image.

Polymer: A huge molecule made up of repeated sequences of smaller molecules called monomers.

Polymerization: A chemical process in which individual monomer molecules bind together into a polymer chain.

Power: The energy to perform work.

Powerhouse: A section of a power plant where the turbines that generate electricity are housed.

Prism: A transparent object with nonparallel sides used to disperse a wave of light.

RAM (random-access memory): Memory space in a computer where the user can alter and temporarily store information.

Rhodopsin: A purple pigment used by the eye's rods to see in the dark.

ROM (read-only memory): Memory space on a computer for permanently storing information and programs that cannot be modified by the user.

Semipermeable: Having pores or openings that allow passage of small molecules but not large molecules.

Shuttle: A device used to carry a bobbin of weft back and forth between the warp threads of a loom.

Slag: Impurities left after the smelting of ore.

Superconductivity: The disappearance of electrical resistance in a conductor, especially when the substance is cooled to extremely low temperatures. Such a state allows the passage of a very large electric current and the creation of a strong magnetic field.

Supersonic: Faster than the speed of sound (about 760 mph at sea level).

Temper: To soften glass at high temperature and cool its surface quickly. Exposing the surface to such compressive stress produces very strong glass.

Tension: A force that stretches or lengthens the members of a structure; opposite of compression.

Thermoplastic: A polymer made up of individual chains that can be melted and reformed.

Thermosetting plastic: A polymer chain that undergoes a chemical reaction when heated. Called cross-linking, the reaction binds polymer chains together.

Thrust: A force that moves an object forward.

Tokamak: A generator with a doughnut-shaped core used in nuclear fusion.

Tomography: A technique for reassembling x-ray information by computer to obtain a three-dimensional image of an object's internal structure.

Torque: A force that produces rotation or twisting.

Transformer: A device that alters voltage.

Transistor: An electronic circuit that is composed of semiconducting material.

Transponder: An electronic transmitting-receiving device that can find, identify, and precisely locate an object or vehicle.

Turbine: A machine with a set of blades mounted on a central shaft rotated by air, water, or steam.

Voltage: The electromotive force, measured in volts, with which a source of electricity sends electrons along a circuit to form a current.

Warp: A set of yarns forming the lengthwise element of a fabric.

Weft: A filling thread or yarn interlaced crosswise with the warp to create a fabric.

Wind load: The force on a structure caused by wind.

Woof: Another term for weft.

Worm gear: A gear of a worm (a screw or other spiral device) that intermeshes with a gearwheel.

Wrought iron: A tough, malleable, commercial iron that has a lower carbon content than steel or cast iron.

X-ray: Electromagnetic radiation with a very short wavelength. It can be detected by phosphor screens and photographic film.

INDEX

HOW THINGS WORK:
EVERYDAY TECHNOLOGY EXPLAINED
By John Langone

Published by the National Geographic Society
John M. Fahey, Jr., *President and Chief Executive Officer*
Gilbert M. Grosvenor, *Chairman of the Board*
Nina D. Hoffman, *Executive Vice President;*
President, Books Publishing Group

Prepared by the Book Division
Kevin Mulroy, *Senior Vice President and Publisher*
Leah Bendavid-Val, *Director of Photography Publishing*
and Illustrations
Marianne R. Koszorus, *Director of Design*

Barbara Brownell Grogan, *Executive Editor*
Elizabeth Newhouse, *Director of Travel Publishing*
Carl Mehler, *Director of Maps*

Staff for this Edition
Susan Tyler Hitchcock, *Project and Text Editor*
Jane Sunderland, *Contributing Project Manager*
Jane Menyawi, *Illustrations Editor*
Jennifer Frink, *Art Director*
Peggy Archambault, *Contributing Art Director*
Margo Browning, *Contributing Editor*
Hal Hellman, *Science and Technology Consultant*
Meredith Wilcox, *Illustrations Specialist*
Connie Binder, *Indexer*
Richard S. Wain, *Production Project Manager*

Rebecca Hinds, *Managing Editor*
Gary Colbert, *Production Director*

Manufacturing and Quality Control
Christopher A. Liedel, *Chief Financial Officer*
Phillip L. Schlosser, *Managing Director*
John T. Dunn, *Technical Director*
Vincent P. Ryan, *Director*
Chris Brown, *Director*
Maryclare Tracy, *Manager*

Founded in 1888, the National Geographic Society is one of the largest nonprofit scientific and educational organizations in the world. It reaches more than 285 million people worldwide each month through its official journal, NATIONAL GEOGRAPHIC, and its four other magazines; the National Geographic Channel; television documentaries; radio programs; films; books; videos and DVDs; maps; and interactive media. National Geographic has funded more than 8,000 scientific research projects and supports an education program combating geographic illiteracy.

For more information, please call 1-800-NGS LINE (647-5463) or write to the following address:

National Geographic Society
1145 17th Street N.W.
Washington, D.C. 20036-4688 U.S.A.

Visit us online at www.nationalgeographic.com.

The Library of Congress has cataloged the earlier edition of this book as follows:

Langone, John, 1929-2006
  The new how things work: everyday technology explained / John Langone ; art by Pete Samek, Andy Christie, and Bryan Christie.
      p. cm.
  Rev. ed. of: National Geographic's how things work. 2004.
  Includes index.
  ISBN 0-7922-6956-X
  1. Technology--Popular works. 2. Inventions--Popular works. I. Title: How things work. II. Langone, John, 1929- National Geographic's how things work. III. Title.

T47.L2923 2004
600--dc22

                                                    2004050438

ISBN 0-7922-5569-0; 978-0-7922-5569-7
ISBN 0-7922-5570-4; 978-0-7922-5570-3 (deluxe)